NATIONAL ACADEMIES
Sciences
Engineering
Medicine

NATIONAL ACADEMIES PRESS
Washington, DC

Simplifying Research Regulations and Policies

Optimizing American Science

Alan Leshner, Alex Helman,
André Porter, and Katie Wullert,
Editors

Committee on Improving
Regulatory Efficiency and Reducing
Administrative Workload to
Strengthen Competitiveness and
Productivity of U.S. Research

Committee on Science, Engineering,
Medicine, and Public Policy

Policy and Global Affairs

Consensus Study Report

NATIONAL ACADEMIES PRESS 500 Fifth Street, NW Washington, DC 20001

This study was supported by contracts between the National Academy of Sciences and the Ralph J. Cicerone and Carol M. Cicerone Endowment for NAS Missions, and the Simons Foundation International. Any opinions, findings, conclusions, or recommendations expressed in this publication do not necessarily reflect the views of any organization or agency that provided support for the project.

International Standard Book Number-13: 978-0-309-99579-5
Digital Object Identifier: https://doi.org/10.17226/29231
Library of Congress Control Number: 2025947557

This publication is available from the National Academies Press, 500 Fifth Street, NW, Keck 360, Washington, DC 20001; (800) 624-6242; https://nap.nationalacademies.org.

The manufacturer's authorized representative in the European Union for product safety is Authorised Rep Compliance Ltd., Ground Floor, 71 Lower Baggot Street, Dublin D02 P593 Ireland; www.arccompliance.com.

Copyright 2025 by the National Academy of Sciences. National Academies of Sciences, Engineering, and Medicine and National Academies Press and the graphical logos for each are all trademarks of the National Academy of Sciences. All rights reserved.

Printed in the United States of America.

Suggested citation: National Academies of Sciences, Engineering, and Medicine. 2025. *Simplifying Research Regulations and Policies: Optimizing American Science*. Washington, DC: National Academies Press. https://doi.org/10.17226/29231.

The **National Academy of Sciences** was established in 1863 by an Act of Congress, signed by President Lincoln, as a private, nongovernmental institution to advise the nation on issues related to science and technology. Members are elected by their peers for outstanding contributions to research. Dr. Marcia McNutt is president.

The **National Academy of Engineering** was established in 1964 under the charter of the National Academy of Sciences to bring the practices of engineering to advising the nation. Members are elected by their peers for extraordinary contributions to engineering. Dr. Tsu-Jae Liu is president.

The **National Academy of Medicine** (formerly the Institute of Medicine) was established in 1970 under the charter of the National Academy of Sciences to advise the nation on medical and health issues. Members are elected by their peers for distinguished contributions to medicine and health. Dr. Victor J. Dzau is president.

The three Academies work together as the **National Academies of Sciences, Engineering, and Medicine** to provide independent, objective analysis and advice to the nation and conduct other activities to solve complex problems and inform public policy decisions. The National Academies also encourage education and research, recognize outstanding contributions to knowledge, and increase public understanding in matters of science, engineering, and medicine.

Learn more about the National Academies of Sciences, Engineering, and Medicine at **www.nationalacademies.org**.

Consensus Study Reports published by the National Academies of Sciences, Engineering, and Medicine document the evidence-based consensus on the study's statement of task by an authoring committee of experts. Reports typically include findings, conclusions, and recommendations based on information gathered by the committee and the committee's deliberations. Each report has been subjected to a rigorous and independent peer review process and it represents the position of the National Academies on the statement of task.

Proceedings published by the National Academies of Sciences, Engineering, and Medicine chronicle the presentations and discussions at a workshop, symposium, or other event convened by the National Academies. The statements and opinions contained in proceedings are those of the participants and are not endorsed by other participants, the planning committee, or the National Academies.

Rapid Expert Consultations published by the National Academies of Sciences, Engineering, and Medicine are authored by subject-matter experts on narrowly focused topics that can be supported by a body of evidence. The discussions contained in rapid expert consultations are considered those of the authors and do not contain policy recommendations. Rapid expert consultations are reviewed by the institution before release.

For information about other products and activities of the National Academies, please visit www.nationalacademies.org/about/whatwedo.

COMMITTEE MEMBERS

DAVID SKORTON (*Chair until May 22, 2025*), Association of American Medical Colleges
ALAN LESHNER (*Chair from May 22, 2025*), American Association for the Advancement of Science (*retired*)
DAVID APATOFF, Arnold & Porter LLP (*retired*)
LINDA COLEMAN, Stanford University
KELVIN DROEGEMEIER, University of Illinois Urbana-Champaign
MELANIE L. GRAHAM, University of Minnesota
LISA NICHOLS, University of Notre Dame
JULIA PHILLIPS, Sandia National Laboratories (*retired*)
STACY PRITT, The Texas A&M University System
STUART SHAPIRO, Rutgers University
CHRISTOPHER VIGGIANI, Oregon State University
EMANUEL WADDELL, North Carolina Agricultural and Technical State University
STEPHEN WILLARD, ICaPath, Inc.

Study Staff

ALEX HELMAN, Study Director
ANDRÉ PORTER, Senior Program Officer
KATIE WULLERT, Program Officer
JOHN VERAS, Associate Program Officer
EMILY McDOWELL, Research Associate
ANDREA DALAGAN, Senior Program Assistant
JORDAN GRAVES, Program Coordinator
TOM WANG, Senior Board Director
RIAN LUND DAHLBERG, Board Director
EMMA COSTA, Mirzayan Fellow

Consultant

JOE ALPER, Science Writer

Reviewers

This Consensus Study Report was reviewed in draft form by individuals chosen for their diverse perspectives and technical expertise. The purpose of this independent review is to provide candid and critical comments that will assist the National Academies of Sciences, Engineering, and Medicine in making each published report as sound as possible and to ensure that it meets the institutional standards for quality, objectivity, evidence, and responsiveness to the study charge. The review comments and draft manuscript remain confidential to protect the integrity of the deliberative process.

We thank the following individuals for their review of this report:

B. TAYLOR BENNETT, National Association for Biomedical Research
KEVIN GAMACHE, The Texas A&M University System
C.K. GUNSALUS, University of Illinois Urbana-Champaign
BRUCE MORGAN, University of California, Irvine
ELIZABETH PELOSO, University of Pennsylvania
KAREN PLAUT, Purdue University
CRAIG REYNOLDS, Van Andel Institute
JOHN ROSENTHALL, Tougaloo College Research and Development Foundation
MEGAN SINGLETON, Johns Hopkins University

JOANNE TORNOW, National Science Foundation
RUTH WILLIAMS, University of California, San Diego

Although the reviewers listed above provided many constructive comments and suggestions, they were not asked to endorse the conclusions or recommendations of this report nor did they see the final draft before its release. The review of this report was overseen by **DAVID ALLISON,** Baylor College of Medicine, and **CHERRY MURRAY,** Harvard University. They were responsible for making certain that an independent examination of this report was carried out in accordance with the standards of the National Academies and that all review comments were carefully considered. Responsibility for the final content rests entirely with the authoring committee and the National Academies.

Acknowledgments

The committee would like to express its gratitude to the many individuals and organizations that made this report possible. First, the committee would like to thank the Ralph J. Cicerone and Carol M. Cicerone Endowment for NAS Missions and the Simons Foundation International for their sponsorship of this study.

The committee is deeply grateful for the many individuals who took time to contribute to the study process and inform the committee's work by providing public testimony, submitting written comments to the request for information, answering questions, and helping the group to understand the scope of the issue and the needs for reform. Although space does not permit identifying them all by name, the committee could not have done its work without their critical contributions.

Finally, this report is only possible thanks to the dedication of the staff at the National Academies of Sciences, Engineering, and Medicine. The study team of Alex Helman, André Porter, Katie Wullert, John Veras, Emily McDowell, Andrea Dalagan, Rian Lund Dahlberg, Tom Wang, and Jordan Graves worked tirelessly to help the committee produce a report on such a short timeline. The committee also extends its thanks to Joe Alper for his writing and editing contributions throughout its work.

Contents

Preface — xv

Acronyms and Abbreviations — xix

Executive Summary — 1

1 Introduction and Context — 5
CURRENT SYSTEMIC CHALLENGES, 6
STATE AND INSTITUTION-LEVEL REQUIREMENTS, 13
MANY RECOMMENDATIONS FOR CHANGE, 14
THE CURRENT PUSH TO REDUCE REGULATIONS, 16
COMMITTEE APPROACH TO THE REPORT, 19
REDUCING BURDEN AND OPTIMIZING SCIENCE, 20
REFERENCES, 23

2 Options to Optimize the Research Enterprise — 27
A NEW APPROACH TO RESEARCH REGULATIONS AND REQUIREMENTS, 28
SYSTEM-WIDE CHANGE, 31
REGULATORY AREA 1: GRANT PROPOSALS AND MANAGEMENT, 36
REGULATORY AREA 2: RESEARCH MISCONDUCT, 43

REGULATORY AREA 3: FINANCIAL CONFLICT OF
  INTEREST IN RESEARCH, 50
REGULATORY AREA 4: PROTECTING RESEARCH
  ASSETS, 54
REGULATORY AREA 5: RESEARCH INVOLVING
  BIOLOGICAL AGENTS, 68
REGULATORY AREA 6: HUMAN SUBJECTS
  RESEARCH, 73
REGULATORY AREA 7: RESEARCH USING
  NONHUMAN ANIMALS, 91
CONCLUDING THOUGHTS, 100
REFERENCES, 104

Appendix A Public Meeting Agendas   **109**
Appendix B Committee Biographical Sketches   **113**

Boxes, Figure, and Tables

BOXES

1-1 Statement of Task, 18
1-2 Illustrative Case of the Current Regulatory Environment, 21

2-1 Illustrative Case of a Possible Future in the Regulatory Environment, 102

FIGURE

1-1 Regulations and policies adopted or substantially modified and changes in interpretation affecting federal research, 9

TABLES

2-1 Options to Address Insufficiently Centralized U.S. Government Oversight of the Regulatory Environment, 33
2-2 Options to Address the Burdensome Grant Processes (Regulatory Area 1), 39
2-3 Options to Address the Problem of Different Standards for Research Misconduct Proceedings Across Agencies (Regulatory Area 2), 45

xiii

2-4	Options to Address the Problem of Slow and Ineffective Digital Infrastructures for Handling Misconduct Cases (Regulatory Area 2),	48
2-5	Option to Address the Uncertain Impact of New HHS Guidelines (Regulatory Area 2),	49
2-6	Options to Address the Inconsistent FCOI in Research Procedures (Regulatory Area 3),	52
2-7	Options to Address Research Security Compliance Issues (Regulatory Area 4),	57
2-8	Options to Address Export Controls (Regulatory Area 4),	63
2-9	Options to Address Cybersecurity and Data Management (Regulatory Area 4),	66
2-10	Options to Address the Complex and Overlapping Regulations for Research Involving Biological Agents (Regulatory Area 5),	70
2-11	Options to Address the Continued Agency Variation in Human Subjects Regulations (Regulatory Area 6),	77
2-12	Options to Address Challenges with Implementing a Single IRB (Regulatory Area 6),	85
2-13	Options to Address the Limited Flexibility and Timeliness Within Existing Regulatory Frameworks (Regulatory Area 6),	87
2-14	Option to Address the Inadequate Adaptation to Evolving Research Methods and Technologies (Regulatory Area 6),	88
2-15	Options to Address the Lack of Federal Guidance on Integrating Nonhuman Subjects Requirements Within Human Research Protection Programs (Regulatory Area 6),	90
2-16	Options to Address the Lack of Harmonization in the Regulation of Research Using Nonhuman Animals Across Federal Agencies (Regulatory Area 7),	94
2-17	Options to Address Burdensome NIH OLAW Requirements (Regulatory Area 7),	96
2-18	Options to Address the Lack of a Sustainable Mechanism for Revising the Guide for the Care and Use of Laboratory Animals (Regulatory Area 7),	99

Preface

The world needs a strong science enterprise now more than at any time in history. Virtually every major issue confronting society has a scientific component to it, either as a cause or a cure, and America has long been among the best in the world at using science to tackle important issues. Maintaining that global eminence has required substantial public trust and financial investment that has been reliable and stable since World War II, and it has paid off handsomely for the United States. Appropriately, along with that trust and investment have come obligations on the part of the scientific enterprise to be transparent in accounting for the responsible use of the funds it receives and to ensure the work is conducted at the highest level of integrity.

However, as the number of federal agencies that support scientific research has grown, and as the science evolves to require new areas of oversight, federal requirements have proliferated, and the workload for researchers, their institutions, and the agencies that fund the research projects have increased to the point of being nearly unmanageable. If this proliferation were only the result of increasing need for oversight, it would be warranted. Instead, federal requirements have become more complex, duplicative, and even contradictory in ways that lead to a more limited gain in productivity and a heavy tax on the time researchers can devote to science. According to the Federal Demonstration Partnership, more than 40 percent of a scientist's research time is now spent on administrative requirements. There is widespread agreement that this proportion is too high, is inhibiting the progress of science, and is therefore limiting return on public investment and benefits to society.

This is not a new problem, and we know much about how to solve it. Conversations about administrative workload were happening when I was a faculty member more than 40 years ago. The issue followed me to my role as an institute director at the National Institutes of Health in the 1990s, when our grantees, who needed to submit proposals to multiple agencies to ensure support in the face of tight funds, had to waste research time rewriting their proposals to meet the idiosyncratic format preferences of each agency. It was still an important issue when I was CEO of the American Association for the Advancement of Science, and I even wrote an editorial calling for a reduction in administrative burden in 2008.[1] Many organizations have weighed in on this issue since, yet it is telling that I am chairing a study on this same topic in retirement. Few recommendations from previous reports have been implemented, and only a small amount of progress has been made reducing the administrative overload for the nation's scientific enterprise.

It is true that the societal context for science has become more complex over the years, and there are more issues that both need and deserve attention. But those changes cannot account for all the added rules, policies, and reporting requirements. There is clear agreement from all the existing analyses that some administrative requirements are overly detailed and that there is too much variation in the ways different agencies approach the same concerns. Much could be accomplished by streamlining reports to agencies, developing common formats for proposals and reports, reducing unnecessary redundancies, and adopting a philosophy of regulating only when there is clear risk to prevent. As I wrote in 2008, "an ideal goal would be for every science-related rule or regulation to be rationalized and streamlined. As a group, they should be integrated as much as possible to reduce unnecessary duplication."[1]

This report takes a different approach from earlier National Academies of Sciences, Engineering, and Medicine studies and others on this issue. Like other analyses, it first identifies the major problem areas requiring policy or regulation reform, but then, rather than making a single recommendation about how to deal with each problem, we offer alternative ways to approach the issue. Any of the options chosen would result in significant progress in reducing the administrative workload on researchers and their

---

[1] Leshner, A. I. 2008. Reduce administrative burden. *Science* 322(5908):1609.

institutions, as well as on the agencies that fund them. We hope this menu will help agencies more easily find the right approaches for them and will stimulate action—something that is challenging but also of tantamount importance at a time of significant reductions to the federal workforce. Critically important, those efforts must be well coordinated across agencies, or the resulting redundancies will offset any progress.

Our nation has never needed the science enterprise to operate at full steam as much as it does now for our health, security, and prosperity. The time is right for streamlining the rules, policies, and requirements that keep that from happening.

During its work, the committee has consulted with numerous individuals and organizations who enthusiastically contributed their thoughts and ideas. We are grateful to them all. We also benefited greatly from the work of the superb and expert staff of the National Academies involved in this project. The project would not have been completed without their efforts, expertise, and wisdom.

Alan I. Leshner
Chair, Committee on Improving Regulatory
Efficiency and Reducing Administrative
Workload to Strengthen Competitiveness and
Productivity of U.S. Research
September 2025

Acronyms and Abbreviations

ACURO	Animal Care and Use Review Office
AI	artificial intelligence
API	application programming interface
BMBL	Biosafety in Microbiological and Biomedical Laboratories
CCL	Commerce Control List
CDC	Centers for Disease Control and Prevention
CHIPS	Creating Helpful Incentives to Produce Semiconductors for America
COC	conflict of commitment
COGR	Council on Governmental Relations
COI	conflict of interest
CUI	Controlled Unclassified Information
DOD	Department of Defense
DOE	Department of Energy
DURC	dual-use research of concern
EO	Executive Order

FCOI	financial conflict of interest
FDA	Food and Drug Administration
FDP	Federal Demonstration Partnership
GAO	Government Accountability Office
GOF	gain of function
HHS	Department of Health and Human Services
HIPAA	Health Insurance Portability and Accountability Act
HRPP	Human Research Protection Program
IBC	Institutional Biosafety Committee
IRB	Institutional Review Board
LOI	letter of intent
NIH	National Institutes of Health
NIST	National Institute of Standards and Technology
NOFO	Notice of Funding Opportunity
NSDD	National Security Decision Directive
NSF	National Science Foundation
NSPM-33	National Security Presidential Memorandum-33
NSTC	National Science and Technology Council
OFAC	Office of Foreign Assets Control
OIRA	Office of Information and Regulatory Affairs
OLAW	Office of Laboratory Animal Welfare
OMB	Office of Management and Budget
ORI	Office of Research Integrity
OSTP	Office of Science and Technology Policy
PHI	protected health information
PHS	Public Health Service
PID	persistent identifier
SciENcv	Science Experts Network Curriculum Vitae
SECURE	Safeguarding the Entire Community in the U.S. Research Ecosystem
SFI	significant financial interest

ACRONYMS AND ABBREVIATIONS

USDA	Department of Agriculture
USG	U.S. Government
USML	United States Munitions List
VA	Department of Veterans Affairs

Executive Summary

The U.S. scientific enterprise has produced countless discoveries that have led to significant advances in technology, health, security, safety, and economic prosperity. However, concern exists that excessive, uncoordinated, and duplicative policies and regulations surrounding research are hampering progress and jeopardizing American scientific competitiveness. Estimates suggest the typical U.S. academic researcher spends more than 40 percent of their federally funded research time on administrative and regulatory matters, wasting intellectual capacity and taxpayer dollars. Although administrative and regulatory compliance work can be vital aspects of research, the time spent by researchers on such activities continues to increase because of a dramatic rise in regulations, policies, and requirements over time.

This is not a new problem. Although numerous studies and reports over the past decades have recommended solutions, there has been little progress in addressing this problem. However, increasing global competition in scientific discovery and innovation, coupled with a national priority to reduce unnecessary regulatory burdens, has set the stage for modifying current administrative and regulatory policies to better ensure that the research community is maximally productive while simultaneously ensuring the safety, accountability, security, and ethical conduct of publicly funded research.

In response to this imperative, the National Academies of Sciences, Engineering, and Medicine convened a committee to conduct an expedited study to examine federal research regulations and identify ways to improve regulatory processes and administrative tasks, reduce or eliminate unnecessary work, and modify and remove policies and regulations that have outlived their purpose while maintaining necessary and appropriate integrity, accountability, and oversight. In most cases, no single best way to achieve these goals exists; therefore, the committee offers options in each area of research regulation rather than specific recommendations. Although the committee did not tier the options, some options presented may be implemented quickly and acted on expeditiously, whereas others require congressional action and may take time. These considerations are noted in the pros and cons of each option.

Ultimately, the committee examined system-wide changes needed to address this problem in seven areas of academic research regulation:

- Grant Proposals and Management
- Research Misconduct
- Financial Conflict of Interest in Research
- Protecting Research Assets, which includes
    - Research Security
    - Export Controls
    - Cybersecurity and Data Management
- Research Involving Biological Agents
- Human Subjects Research
- Research Using Nonhuman Animals

For each area of regulation, the committee outlines the key problems researchers face navigating the current regulatory environment and provides a table detailing potential options for response. Within each table, the committee presents the goal of the option, the approach to implement it, and the pros and cons to consider before implementation (see the detailed options in the corresponding sections of Chapter 2).

In total, the committee proposes 53 options. By addressing these critical challenges, the committee's report provides a roadmap for establishing a more agile and resource-effective regulatory framework for federally funded research. Such a framework can liberate researchers from unnecessary administrative tasks, empower them to focus more on conducting research and training the next generation of scientists and engineers, and enable U.S. science and technology to thrive, unencumbered by unnecessary bureaucratic obstacles that rob the nation's research enterprise of time and money.

The committee believes that progress can be made by implementing any of the options within a given area. However, the committee also identifies three overarching principles to guide future decision-making:

- Harmonize regulations and requirements across federal and state agencies and research institutions. This may require compromising in the name of harmonization on the type, specificity, and format of information that a given agency requests.
- Take an approach tiered to the nature, likelihood, and potential consequences of risks for a new regulation or requirement. Increased oversight may be needed for higher-risk activities, but more flexibility should be allowed for projects less likely to present risks.
- Use technology to simplify the process of complying with regulations and requirements to the greatest extent possible, following the proven example of the financial industry in using artificial intelligence and machine learning to facilitate compliance activities.

Although the options outlined in this report will take varying degrees of effort and resources to implement, examples of recent reform success stories could serve as models of implementation and demonstrate the value of embracing new approaches. For example, federal agencies, working within the structure of the National Science and Technology Council and with oversight from the Office of Science and Technology Policy, developed coordinated research security policy, forms, guidance, and definitions to the extent possible and continue to work together on common agency implementation.

As administrative requirements adapt to the growing challenges faced by the research enterprise, the committee encourages policymakers to consider the three principles outlined above and in Chapter 2 of this report when adopting any new policies or approaches. The committee calls on all participants in the research enterprise to engage in this potentially transformative effort deliberately and thoughtfully, with energy, urgency, and a mindset that prioritizes results while preventing irreparable harm to U.S research. By doing so, the U.S. regulatory enterprise can accomplish its mission of ensuring that federally funded research is safe, is conducted with integrity, maximizes the value of taxpayer dollars, and protects the interests of the public without unnecessarily burdening the U.S. research ecosystem and inhibiting its contributions to national well-being, prosperity, and security.

1

Introduction and Context

The American scientific, engineering, and biomedical enterprise has long been viewed as among the best in the world, and researchers have come from almost every country to study and work in the United States. Although U.S. scientific efforts have yielded significant advances in technology, health, security, safety, and prosperity, there is the concern that excessive, uncoordinated, duplicative, and inconsistent policies and regulations are hampering progress in science.

It is imperative that scientific research conforms to the highest professional standards. Not only do high standards promote good stewardship of taxpayer dollars by ensuring accountability, transparency, ethical conduct, security, and safety, but they also promote good science that is rigorous, reliable, and reproducible (NASEM, 2016; OSTP, 2025). Appropriate research regulations and oversight play an important role in promoting research excellence, but the current regulatory system also produces many unintended consequences that can hinder research and place undue and costly burdens on the research enterprise, which have increased as regulations have proliferated over the past several decades.

Unfortunately, federal regulations, policies, requirements, and requested reports[1] have grown to such an extent that they can encumber

---

[1] The committee acknowledges that some states have also added regulations, policies, and requested reports that further add to the burden imposed on researchers and their institutions. However, addressing state regulatory activities is outside the scope of this report's Statement of Task (see Box 1-1).

the research enterprise, hinder innovation, and divert time, resources, and expertise away from research and toward administrative tasks that do not directly benefit research subjects or enhance research outcomes (FDP, 2020; NASEM, 2016). This reduces both research productivity and the time available for training and educating the next generation of investigators (NASEM, 2016). Therefore, within appropriate bounds for accountability and oversight, improving regulatory efficiency and reducing administrative workload are critical goals for fostering innovation and productivity in the U.S. research enterprise. As the burden of complying with increasingly complex regulations and administrative requirements can now outweigh their intended benefits, identifying strategies to streamline operations has become crucial to ensuring more of our nation's investments in scientific discoveries are directed toward research, not "bureaucratic box checking," as Michael Kratsios, director of the White House Office of Science and Technology Policy, noted in a recent address to the National Academies of Sciences, Engineering, and Medicine (The White House, 2025a).

CURRENT SYSTEMIC CHALLENGES

The Cost of Regulations

In today's world, global competitiveness in science is imperative and tied intrinsically to economic, military, and health leadership. However, estimates by the Federal Demonstration Partnership (FDP) suggest the typical academic researcher in the United States spends more than 40 percent of their research time on administrative and regulatory matters rather than actually conducting research (FDP, 2020; Rockwell, 2009). FDP surveys also show that satisfying growing regulatory demands is challenging research institutions and requiring them to use diminishing resources to hire more staff. The United States needs to address this long-standing issue so that researchers can spend their time and resources more effectively on generating the scientific advances that have powered the nation's economy for the past eight decades.

In addition to the time spent complying with administrative and regulatory requirements, there is a monetary cost involved. Data from the Fiscal Year 2020 U.S. National Science Foundation (NSF) Higher Education Research and Development Survey showed that the 116 U.S. institutions receiving more than $100 million in federal research funds in 2020 estimate spending an average of $444,008 in 1 year of complying with research security disclosure requirements; institutions receiving less than $100 million a year estimate spending an average of $100,202 annually (COGR, 2022; NCSES, 2021). A 2022 survey by the Council on Governmental Relations (COGR) found the cost of complying with the National Institutes of Health (NIH) Policy for Data Management and Sharing was estimated to be nearly $1.4 million a year for institutions receiving more than $100 million in federal research funds and just over $1 million a year for smaller institutions (COGR, 2023). Addressing this issue can reduce overhead costs at research institutions at a time when they are operating under tightening budgets.

Researchers, administrators, and compliance officers at academic institutions are not the only ones affected by an increasing administrative workload. Federal agency staff who oversee and manage research funding also face growing administrative demands. Consequently, addressing and reducing burden is all the more imperative in the current context of a reduced federal workforce—and particularly a reduced federal research funding workforce. While the committee's task was to focus on reducing administrative workload for researchers, efforts to streamline, harmonize, modernize, and reduce duplicative requirements will, even with certain upfront costs in implementing change, ultimately serve all parts of the research ecosystem.

Regulatory Implementation

Some of the most significant regulatory challenges stem from concurrent and disparate implementation of regulations and requirements across different federal funding agencies, as well as differences in oversight implementation (GAO, 2016). This lack of harmonization across agencies gives rise to issues that include duplicative efforts, inconsistencies, and even direct contradictions in requirements, all of which consume time that would otherwise be devoted to conducting research.

The current system encompasses variations in how federal regulations affect different agencies, as well as in agency-level policies, reporting requirements, proposal submission processes, conflict of interest requirements, and training (COGR, 2025; NSF and NSB, 2014). For example, two main agencies regulate animal research—the U.S. Department of Agriculture (USDA) under the Animal Welfare Act and Animal Welfare Regulations, while the Office of Laboratory Animal Welfare (OLAW) maintains statutory authority under the Health Research Extension Act, which incorporated into law the Public Health Service (PHS) Policy that OLAW interprets. In some cases, USDA and PHS requirements may conflict with one another. Depending on the circumstances, research may be subject to both sets of requirements, leading to confusion, redundancy, and extra work (NIH OLAW, 2024). In addition, past reports have argued that when requirements are inconsistent or duplicative, the natural result is for academic institutions to create additional requirements of their own to manage the complexity and risk of noncompliance stemming from regulatory complexity. As noted in the next section, adding to the complexity and increased administrative work are additional regulations and requirements enacted by some states and even the institutions themselves (NASEM, 2016).

Finally, research regulations and requirements can be unduly burdensome for researchers when the requirements and systems needed to remain compliant are difficult to navigate. Federal policies and requirements are not always updated sufficiently to reflect changes in the way research is done or the impact of new technologies and consequently do not meet the needs of either researchers or those with oversight responsibilities (NASEM, 2009a, 2025). Forms and systems used to share necessary data or information may be challenging to navigate, vary across agencies, or require multiple systems to perform one similar task, such as preparing annual reports and various registrations (GAO, 2016). Researchers are therefore left with additional work peripheral to their scientific training to determine how potentially out-of-date guidelines apply to their current work or how to manage confusing paperwork or technology systems.

Growth of Regulations

Rather than moving toward a more streamlined approach, however, the administrative workload has increased during the past decade, with added regulations and requirements consuming even more time that should be—and used to be—dedicated to conducting research. As Figure 1-1 shows, 62 percent of regulations and policies that have been adopted or substantially modified and changed since 1991 were issued between 2014 and 2024.

In 2016, the National Academies, at the request of Congress, issued a report, *Optimizing the Nation's Investment in Academic Research: A New Regulatory Framework for the 21st Century*. This report concluded that "the continuing expansion of federal regulations and requirements is diminishing the effectiveness of the U.S. research enterprise and lowering the return on the federal investment in basic and applied research by diverting inves-

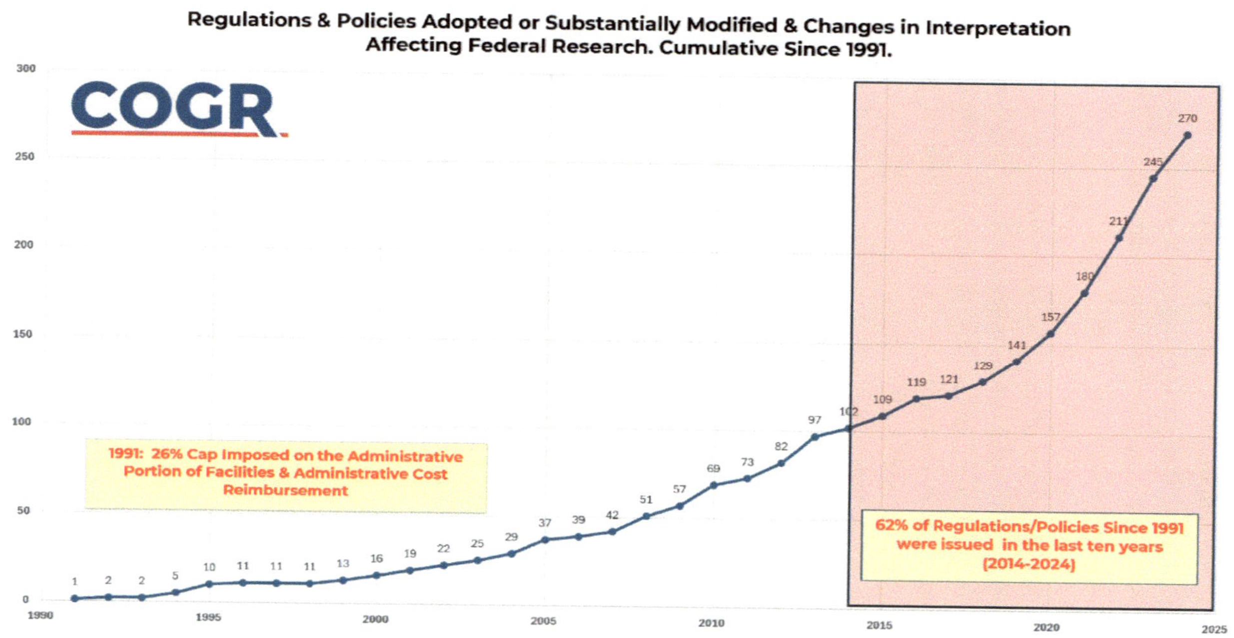

FIGURE 1-1 Regulations and policies adopted or substantially modified and changes in interpretation affecting federal research.
SOURCE: Presented to the committee by Matt Owens on May 21, 2025.

tigators' time and institutional resources away from research and toward administrative and compliance matters," while also acknowledging that "effective regulation is essential to the overall health of the research enterprise." To address this problem, the committee who authored the report recommended steps to improve regulatory efficiency and reduce administrative workload for the nation's academic research enterprise (NASEM, 2016). However, while a few recommendations were adopted, such as employing a single Institutional Review Board (IRB) for studies involving multiple institutions, many of the 2016 report's recommendations remain only partially addressed, and in 2025, compliance burdens placed on researchers remain a significant concern while regulatory burden has continued to compound (CRS, 2017; FDP, 2020; GAO, 2021).

For example, regulations continue to proliferate in the realm of research security. National Security Presidential Memorandum-33 (NSPM-33), issued January 2021, instituted broad requirements for disclosure and established a research security infrastructure focused on cybersecurity, foreign travel security, research security training, and export control training (NSTC, 2022) for recipients of federal research and development funds that exceed $50 million annually. NSPM-33 directed federal agencies to create common forms for the disclosure of foreign affiliations, appointments, and funding sources. Several agencies have made research security resources available. In 2024, the NSF released the Trusted Research Using Safeguards and Transparency framework to help with institutional evaluations of risk related to foreign ties (NSF, 2024). This framework includes standardized training modules and resources for institutions and researchers to access and adapt. Working with the research community through cooperative agreements, NSF, NIH, Department of Defense, and Department of Energy made research security training modules available and more recently endorsed a condensed version of the training developed by the Safeguarding the Entire Community in the U.S. Research Ecosystem (SECURE) framework.

Recent government directives have introduced additional compliance requirements, placing significant focus on "conflicts of commitment"—a new concept for institutions when first introduced (COGR, 2021). Conflicts of commitment occur when a researcher dedicates time to personal activities in excess of institutional policy or that may detract from their professional responsibilities (ORI, n.d.). This has led institutions to develop and implement conflict of commitment programs. The ambiguity of some of the requirements has led to variation in how institutions are developing their infrastructure to comply with the requirements and how internal policies affect researchers and trainees at different institutions.

Attempts to regulate Controlled Unclassified Information (CUI) also vary across agencies, and recent requirements for training allow for unequal implementation and standards across and within institutions. While allowing for varied implementation can streamline regulations and reduce burden by ensuring requirements are not unnecessarily strict, other challenges can emerge when requirements are unclear or introduce uncertainty about how they should be applied. Additional and ambiguous requirements have added a significant cost burden for institutions as they reconcile conflicting definitions to develop the infrastructure needed to comply with requirements related to policies governing export control, such as controlled items and restricted party lists, as well as with safeguards for CUI, such as implementing access control and encryptions. Smaller, less resourced institutions are affected disproportionately.

In addition to the relatively new research security requirements, the federal government has an extensive export control regulatory regime, which has long needed regulatory reform, to protect U.S. trade and national security (NASEM, 2009b, 2022). Academic institutions have experienced significant challenges fully adopting existing federal export control framework requirements given the expansive research areas within academia that require a broad knowledge of regulations, compared to industry where organizations focus on a smaller number of technologies. To meet the demands of complying with both export controls and research security regulations and requirements, institutions have had to find ways to identify and coordinate resources needed to increase their efforts for coming into compliance and centralizing activities within the institution (COGR, 2022). They have also had to address the necessary training for personnel across the institution.

As a final example, new regulations for research misconduct from the U.S. Department of Health and Human Services (HHS) were recently enacted[2] (CITI, 2024; ORI, 2024), but it is unclear what effect these may have on the regulatory environment for misconduct. The committee discusses the potential here for efforts to review the effectiveness and efficiency of these revisions on a clearly delineated timeline.

This year, significant federal actions and policy changes, some of which have come quickly without sufficient implementation guidance and consid-

---

[2] *Public Health Service Policies on Research Misconduct,* 42 C.F.R. Part 93 (September 17, 2024).

eration of their effects on the scientific enterprise,[3,4] have added administrative workloads and created uncertainty, especially for research universities (Dorgelo and Leibenluft, 2025; EAB, 2025). As national research priorities and policies are changed, it is important that any actions taken recognize the associated effects on administrative workload and the efficiencies of scientific research—not only for the administrative offices and researchers of the performing organizations but also for federal research sponsors.

Balancing Oversight and Efficiency

In addition to the growth in regulations and requirements and a lack of harmonization and the problems it creates, another challenge is the difficulty balancing needed regulation and efficiency. Accountability, safety, security, and transparency in research are important, but there are times when regulations can be so stringent and inflexible that they unnecessarily regulate lower-risk research activities. Reports from the National Academies, the U.S. Government Accountability Office (GAO), and others have argued that overly stringent regulations that do not provide flexibility for lower-risk situations can increase workload without improving outcomes (COGR, 2017; GAO, 2016; NASEM, 2016). Ultimately, in a setting where regulations are not calibrated to risk, researchers can spend time and effort on compliance for low-risk activities that could be better used for conducting research and giving greater attention to higher-risk work.

Concern about regulatory or administrative workload, however, does not mean that federal oversight of research is inappropriate. With taxpayer funds supporting U.S. scientific research, the enterprise must ensure full transparency and that research adheres to the highest standards of integrity. Furthermore, careful oversight is necessary to ensure the safety of human research participants, welfare of research animals, protection of intellectual property, and safeguarding of the public and the environment. Developed effectively, regulations provide a framework for conducting research that embodies the shared values of the federal government, research institutions, researchers, and the public.

---

[3] *Association of American Universities v. National Science Foundation,* Civil Action Number 1:25-cv-11231-IT (D. Mass. 2025).

[4] *National Association of Diversity Officers in Higher Education v. Donald J. Trump,* Case Number 1:25-cv-00333-ABA (D. Md. 2025).

STATE AND INSTITUTION-LEVEL REQUIREMENTS

Although numerous challenges exist within the current federal regulatory ecosystem, other factors affect the time and resource requirements researchers face. The Statement of Task for this report focused the efforts of the committee on possible actions to be taken at the federal level. However, state governments and academic institutions also play a role in increasing researcher administrative workload. State-level requirements sometimes duplicate or complicate federal regulatory requirements, and while tackling this issue is out of scope for this report, the committee acknowledges that this issue exists and needs to be addressed.

Along with the potential complications of additional state regulations, academic institutions also develop their own policies and processes to ensure compliance. Institutions are often risk-averse and respond to uncertainty in the regulatory environment as well as their own concerns about the potential for noncompliance by interpreting policies in the strictest manner, even when they could apply a more lenient standard and still be compliant (NASEM, 2016). Some institutions, in particular small- to medium-sized institutions, may adopt zero or near-zero risk-tolerance strategies for research compliance, resulting either in extraordinarily burdensome processes that may entail multiple additional layers of institutional bureaucracy or declining to participate in research with even a slight risk (COGR, 2022; Jager, 2023). Institutions have also been hesitant at times to fully adopt new and more streamlined processes because of risk concerns (Burr et al., 2022).

In addition, while all research institutions navigate the compliance of federal regulations, their impact is not equal. Larger and better-resourced institutions dedicate significant administrative infrastructure and personnel to manage growing compliance demands, which can strain even their more substantial resources. The large administrative workforces needed to manage risk and compliance can further increase bureaucracy and time spent navigating internal processes. For smaller institutions, as well as under-resourced institutions that often have less money for research and are more likely to serve disadvantaged and low-income student populations, the consequences can often be more pronounced, as they may lack the personnel, infrastructure, or expertise to adequately address the increasing volume of regulatory and other requirements. In some cases, this prohibits their participation in research for which they are otherwise well qualified.

Ultimately, this report focuses on the federal government. Nonetheless, greater coordination and cooperation across federal, state, and institutional levels would benefit regulatory optimization and limit unnecessary administrative impact on the researcher.[5]

MANY RECOMMENDATIONS FOR CHANGE

In general, reform is a challenging process with many potential barriers. In large institutions, change can be hindered by multiple veto points that can derail action: investing time and resources into systems developed based on past decisions that become baked into institutional structures; cultural and normative barriers; and individual actors pushing against the upset reform can cause (Bannink and Resodihardjo, 2006). Moreover, any efficiencies that are realized have largely been offset by new requirements.

Thus, despite many expert groups calling for change over the past couple of decades, there has been little progress in reducing regulatory burden. For example, the National Academies' 2016 report included four overarching recommendations that formed the basis of its 40 specific recommended actions. First, that committee called on Congress, the administration, federal agencies, the Office of Management and Budget (OMB), and research institutions to take collective action to critically reexamine and recalibrate the regulatory regime. That committee also recommended establishing a new entity, a Research Policy Board, to provide a public-private forum through which to engage the development and harmonization of research regulations. The 21st Century Cures Act, passed in 2016, directed OMB to establish a Research Policy Board, but OMB never created the board, and authority for the board expired in September 2021. In addition, the committee recommended strong action by universities, in partnership with the proposed Research Policy Board, to ensure institutional and individual integrity in scientific research and to hold institutions accountable for failure to uphold such integrity. Finally, the committee recommended adjusting the responsibilities of inspectors general to balance the need to weed out waste, fraud, and abuse with ensuring economy, efficiency, and effectiveness in research.

Other groups, including COGR, FDP, GAO, the Institute for Responsive Government, the Association of American Medical Colleges, the Association of American Universities, and the Association of Public and

---

[5] Though this is out of the scope of this report, one possible model might be the NSF-funded SECURE Center that works with the research community to identify, prioritize, and collaboratively design and develop the research security resources and tools needed.

INTRODUCTION AND CONTEXT   15

Land-grant Universities, have also highlighted the challenges researchers face, with many offering their own recommendations to reduce the regulatory burden (AAMC, 2020; AAU, 2017; APLU, 2024; COGR, 2025; FDP, 2014; GAO, 2016; Institute for Responsive Government, 2025; NSF and NSB, 2014). In 2020, OMB completed a major rulemaking on this subject after reviewing thousands of comments from public and government sources. Commenters reported that "grants managers ... [were] spending a disproportionate amount of time using antiquated processes to monitor compliance" and navigating duplicative and ineffectual requirements.[6]

Most recently, COGR released a set of "Actionable Ideas to Improve Government Efficiency Affecting the Performance of Research" (COGR, 2025). This document addressed 18 topic areas of research regulations, including research project proposal development, financial conflicts of interest, data management and sharing, animal and human subjects research, and cybersecurity. Overarching themes of the COGR recommendations stressed the need to develop single systems and consistent processes across many areas of research regulations, allow risk-tiered variation in regulations, and update and revise critical definitions, including the definitions of clinical trials, gifts, and fundamental research.

There is significant overlap across the recommendations from COGR's 2025 report, the 2016 National Academies' report, and many of the other reports cited above. These reports have called for agencies to adopt harmonized and centralized mechanisms for reporting and oversight and ensure there are usable and up-to-date systems and definitions that provide clarity to researchers, all while more appropriately balancing the need for oversight with ensuring that scientific work remains efficient and effective. The themes in these calls for action align with the greatest challenges of the present regulatory system.

There has been some uptake and implementation of previous recommendations for reducing regulatory burden. In 2024, for example, OMB updated the Uniform Guidance for grants management and financial oversight to increase the audit threshold to $750,000, reducing burden on smaller institutions. In addition, HHS and NIH instituted policies to simplify the peer review process through updates to 2025 grant applications. However, as Chapter 2 discusses in more detail, multiple areas of regulation need similar types of changes to truly address the magnitude of the problem. An improved regulatory environment requires increased harmonization,

---

[6] *Guidance for Grants*, 85 Fed. Reg. 49506 (August 13, 2020).

greater adoption of approaches tiered to risk, and user-friendly technology to simplify processes. These principles are repeated consistently throughout the specific policy options detailed in Chapter 2.

THE CURRENT PUSH TO REDUCE REGULATIONS

Shortly after taking office in 2025, President Trump issued an Executive Order requiring federal agencies to repeal at least 10 existing rules, regulations, or guidance documents for every new rule, regulation, or guidance they propose (The White House, 2025b). As part of this effort, OMB issued a Request for Information soliciting ideas for rules and regulations that could be rescinded to reduce administrative and regulatory burdens.[7] These actions, combined with budget pressure at the federal and institutional levels, the wealth of proposed solutions available to addressing this problem, and the availability of technologies that can shoulder some of the workload, have created an environment in which the current administration is welcoming proposals to increase regulatory efficiency and reduce administrative workload.

For example, the rapid advancement and potential of artificial intelligence (AI) has created opportunities to reduce the time and monetary costs of satisfying regulatory requirements and completing administrative tasks. Proposed uses of AI to reduce administrative tasks include assisting with the preparation of application materials, generating financial reports for grant management, and creating reports needed for regulatory compliance. AI also has the potential to make the use of publicly funded research far more valuable if research reports are presented in a uniform manner, which in turn is more conducive to the use of AI to mine those reports for additional insights.

At the same time, since January 20, 2025, the federal government and federal funding agencies have renewed their interest in improving government efficiency and reducing federal outlays. With federal research agencies facing billions of dollars in budget cuts and reductions in staffing, there is not only the opportunity but also the necessity to optimize the nation's investment in academic research by allocating more time and money to conducting research and reducing the time and money spent on administrative tasks. Several federal agencies have also proposed capping indirect

---

[7] *Request for Information: Deregulation*, 90 Fed. Reg. 15481 (April 11, 2025).

cost reimbursements,[8] though these caps are currently the subject of multiple legal challenges. This environment has made it even more necessary to thoughtfully and carefully consider how to best use researchers' limited time and institutions' limited budgets while preserving an appropriate level of regulatory oversight.

Given the current emphasis on reducing federal discretionary spending and maximizing the return on the funds allocated for research, as well as the well-documented administrative workload currently facing the nation's researchers, the National Academies recognized this as an opportune time to revisit the topic of how to improve regulatory efficiency and reduce the administrative workload imposed on the nation's academic research enterprise with the goal of strengthening the competitiveness and productivity of U.S. research. With the support of the Ralph J. Cicerone and Carole M. Cicerone Endowment for NAS Missions and Simons Foundation International, the National Academies' Committee on Science, Engineering, Medicine, and Public Policy, in collaboration with the Board on Higher Education and Workforce, convened an ad hoc committee to conduct an expedited study to identify strategies and actionable options aimed at streamlining regulatory processes and administrative tasks, reducing or eliminating unnecessary administrative work, and removing policies and regulations that were ill conceived or have outlived their purpose while maintaining necessary and appropriate integrity, accountability, and oversight (see Box 1-1 for the committee's Statement of Task).

By addressing these challenges, the committee's report aims to provide a roadmap for establishing a more agile and resource-effective regulatory framework. Such a framework can liberate researchers from unnecessary administrative tasks, empower them to focus on research and training the next generation, and enable U.S. science and technology to accelerate and thrive. American leadership in science and technology is important, but it is slipping (ASPI, 2024). By leading the scientific enterprise, the United States has control over the outcomes produced and can promote the values that underpin the application of those outcomes, such as the ethical use of AI.

---

[8] Indirect cost reimbursements are the funds research universities use to pay for the facilities, equipment, systems, compliance boards, certifications, and staffing necessary to conduct the research and meet federal regulatory and administrative reporting requirements.

BOX 1-1
Statement of Task

Over the past two decades, questions have continued to arise about the cost of research in the United States, and whether the growing number of federal regulations increase the monetary and time costs to individual researchers and their institutions. Several recent reports have identified ways to reduce the regulatory burden, but many of those recommendations have not been implemented. A committee of the National Academies of Sciences, Engineering, and Medicine will review and prioritize federal actions that could improve regulatory efficiency and potentially reduce costs in the academic research environment, particularly for the academic researcher.

The committee will undertake an expedited effort to describe the impacts of administrative workload and current regulations on research productivity; analyze federal research regulations in light of the 2016 National Academies report *Optimizing the Nation's Investment in Academic Research: A New Regulatory Framework for the 21st Century* to determine whether the report's recommendations for regulatory change have been implemented; and examine other recommendations from reports developed by such groups as the Association of American Universities, Association of Public and Land-grant Universities, and Council on Governmental Relations, and others on the impacts of federal regulations on researcher and institutional workload.

The committee will produce a brief report that presents a menu of prioritized options for federal actions to improve regulatory efficiency affecting researchers and their institutions, including initiatives by the White House and executive agencies or Congress. The options presented will describe the anticipated impacts on reducing different types of administrative workload, noting potential unintended consequences, while minimizing risk to accountability and research performance. Finally, the committee will describe, to the extent possible, new developments, such as the application of new technologies like artificial intelligence, that could improve administrative efficiency.

COMMITTEE APPROACH TO THE REPORT

In this report, the committee outlines a set of steps and options the federal government might take to create a system that better allows researchers to focus on what they do best—conducting scientific research—while also meeting the need to conduct their research responsibly, ethically, and with appropriate stewardship of taxpayer-provided funding. Rather than make explicit recommendations for what the federal government should do to achieve these goals, as previous reports have done, the committee is providing a set of options, from which policymakers and other key actors can weigh pros and cons to choose the best approach to achieve these goals, accompanied by discussions of important factors to consider before undertaking any of the options. Although the committee did not tier the options, some options presented could be acted on expeditiously, whereas others may require congressional action and time. These considerations are noted in the pros and cons of each option. While the scope of the committee's efforts was limited to the academic research enterprise, many of the options presented here could serve to reduce regulatory workload across the U.S. research ecosystem more broadly.

The administrative workload of complying with research requirements is a result of actions originating over time in many sectors, including federal agencies, academic research institutions, and state governments. Therefore, reducing that workload will require all participants to work together to create lasting improvements. However, because the charge to the committee is limited to "federal actions to improve regulatory efficiency," the committee is providing policy options that only apply to the federal government.

Given the urgency to better align regulatory and oversight processes with the need to responsibly maximize quality research, the committee was asked to conduct its work in a compressed 4-month time frame and produce a report that is shorter and therefore less comprehensive than many National Academies' reports. Consequently, the options presented in this report are considered by the committee to provide the greatest potential impact, but they do not represent an exhaustive list of all the actions that could be taken to improve regulatory efficiency. While the options are presented generally in the order of likely effectiveness, the committee acknowledges that these rough estimates are difficult to assess without a more robust evaluation. In many cases, the options presented differ enough from one another that it is challenging to determine which would produce the greatest improvement, and others may argue for a different ordering. Therefore, the committee is not suggesting that options stated earlier are inherently preferred to those stated later. Implementing even a few of these options could go a long way toward improving regulatory efficiency and decreasing administrative workload for both federal agencies and researchers.

The committee also addresses many unique and specific challenges facing different areas of regulation resulting from updates to the regulatory environment, legal changes or provisions, and specific guidance. For example, within human subjects research, adopting a previous National Academies' recommendation to establish a single IRB[9] for studies with multiple sites resulted in unintended consequences (NASEM, 2016). While the intent of this recommendation was to allow multi-institutional projects to use a single IRB process, thereby streamlining review, mandatory adoption has created its own distinct challenges that the committee seeks to address with options that can improve this structure.

This report is organized in two chapters. This chapter provides an overview of the major problems with the current research regulatory framework that require attention, the current situation and push for reducing regulations, and previous approaches to regulatory changes. The second chapter highlights key issues with current regulations on each of several specific topics, potential approaches to and options for addressing those issues, and potential positive and unintended consequences of each approach. It concludes with closing thoughts from the committee and a look to the future of regulating the U.S. scientific enterprise.

REDUCING BURDEN AND OPTIMIZING SCIENCE

This chapter paints a picture of a system operating under a heavy regulatory workload that hinders science without sufficient gains in transparency and integrity. The challenges of the current system are captured in the experiences of researchers navigating this complex landscape: it is in the stories of researchers that it becomes clear how much overly complex and duplicative regulations can cause problems for science as a whole. The committee concludes the chapter with a look into the ways in which excessive regulatory workload hinders science in the day-to-day lives of researchers. As an example, consider the case of Dr. Linh Tran, a hypothetical researcher facing many of the common challenges seen in the current regulatory environment (see Box 1-2).

---

[9] *Cooperative Research*, 45 C.F.R. 46 114(b)(1) (January 19, 2017).

BOX 1-2
Illustrative Case of the Current Regulatory Environment

Dr. Linh Tran, an associate professor of robotics at an emerging research university, is a rising star. When she was young, her sister was in an accident and lost the use of her arm. As a young girl, Linh was convinced that science and medicine could help people like her sister, and she committed herself to science and to developing artificial limbs.

Dr. Tran had made breakthroughs in brain-computer interfaces and was on her way to developing highly functional, lifelike prosthetics. With her new work to incorporate artificial intelligence into her prototypes, she was confident she could develop devices that were more responsive and could ultimately be controlled by an individual's thoughts.

Dr. Tran received the trifecta of funding: a grant from the U.S. National Science Foundation (NSF), a contract from the U.S. Department of Defense (DOD), and a research sponsorship from an established robotics company. Dr. Tran was excited. But she was also exhausted. Simply applying for the funding had taken months. Her ideas were well formulated, but she needed to draft lengthy proposals that included much more than her science. She navigated different NSF and DOD submission portals and processes, often submitting the same information about herself and her team, her funding sources, institution, budgets, plans for compliance, and more, all in different formats for the two funding agencies.

As the projects kicked off, Dr. Tran updated her financial conflict of interest (FCOI) disclosures. The university had developed a single annual disclosure form, but each sponsor had a different definition of what counted as a "significant" financial interest, prompting a number of follow-up questions. Dr. Tran did not need a formal FCOI management plan, but this review prompted an additional research security and export control review, ultimately requiring her to develop a technology control plan, particularly for the advanced actuator that was needed for her prototype to make physical movements. This came with additional training modules for Dr. Tran and her team to complete before research could begin.

One of the postdocs on Dr. Tran's team was a brilliant biosystems engineer from South Korea who sought to test new prototypes on human research participants. Including the postdoc on the DOD project triggered a lengthy foreign national approval

continued

> **BOX 1-2 Continued**
>
> process, a delayed visa approval, and an additional export control review. Despite arriving late, the postdoc was still ready to go. His Institutional Review Board (IRB) protocol was approved in about 4 months, but it needed an additional approval from a DOD IRB, which resulted in small but significant changes in the consent forms, prompting a second approval by the university's IRB and adding further delays.
>
> Dr. Tran felt demoralized. After a year of funding she felt like her team had barely gotten to the actual research.

Dr. Tran's experience is not purely hypothetical, however. As part of its information-gathering efforts, the committee sent out a request for information on May 6, 2025, seeking input from the scientific community on how to improve regulatory efficiency. Within a few weeks, the request had amassed nearly 200 responses with detailed discussions of the ways in which the current system not only frustrates researchers and institutions but also takes away crucial time from scientific discovery. Across many different respondents, the committee heard similar refrains:

> One of our biggest challenges has been navigating the complex and fragmented landscape of federal compliance requirements. . . . The administrative burden of reporting, matching fund requirements, and indirect cost negotiations often *diverts time and personnel from core innovation activities.*

> Some published estimates of administrative burden place this as high as 40% of faculty [research] time . . . so in that context the administrative burden means that *faculty have very little time left for the research itself.*

> One major hurdle is the lack of standardized protocol forms. . . . This lack of uniformity not only creates unnecessary administrative burden but *can also slow down the start of important research projects.* (Request for Information responses, emphasis added)

As these responses highlight, time spent managing overwhelming regulatory variation and complexity is time spent away from scientific discovery and training the next generation. Each duplicative form, uncertain requirement, and outdated reporting system chips away at the work done to advance our understanding of the world around us and diverts critical federal funding away from discovery and innovation.

REFERENCES

AAMC (Association of American Medical Colleges). 2020. Aamc conflicts of interest metrics project - measuring the impact of the public health service regulations on conflicts of interest. https://www.aamc.org/media/50386/download (accessed June 12, 2025).

AAU (Association of American Universities). 2017. Policy recommendations: Research regulatory reform. https://www.aau.edu/sites/default/files/AAU Files/Key Issues/Research Administration %26 Regulation/AAU-Regulatory-Reform-Policy-Recommendations.pdf (accessed June 12, 2025).

APLU (Association of Public & Land-Grant Universities). 2024. Letter to the chairwoman of the energy and commerce committee from aplu. https://www.aplu.org/wp-content/uploads/APLU-EC-NIH-letter-2024-signed.pdf (accessed June 24, 2025).

ASPI (Australian Strategic Policy Institute). 2024. *ASPI's two-decade critical technology tracker: The rewards of long-term research investment.* https://www.aspi.org.au/report/aspis-two-decade-critical-technology-tracker/ (accessed July 22, 2025).

Bannink, D., and S. Resodihardjo. 2006. The myths of reform. In *Reform in Europe*, 1st ed., edited by L. Heyse, S. Resodihardjo, T. Lantink, and B. Lettinga. New York: Routledge. Pp. 1–32.

Burr, J. S., A. Johnson, A. Risenmay, S. Bisping, E. S. Serdoz, W. Coleman, et al. 2022. Demonstration project: Transitioning a research network to new single IRB platforms. *Ethics & Human Research* 44(6): 32–38. https://doi.org/10.1002/eahr.500149.

CITI (Collaborative Institutional Training Initiative). 2024. *HHS finalizes rule on research misconduct to strengthen research integrity.* https://about.citiprogram.org/blog/hhs-finalizes-rule-on-research-misconduct-to-strengthen-research-integrity/ (accessed July 15, 2025).

COGR (Council on Governmental Relations). 2017. *Reforming animal research regulations: Workshop recommendations to reduce regulatory burden.* https://www.cogr.edu/sites/default/files/Animal-Regulatory-Report-October2017.pdf (accessed June 24, 2025).

COGR. 2021. *Principles for evaluating conflict of commitment concerns in academic research.* https://www.cogr.edu/sites/default/files/Final%20for%20publication%20COC%20Principles%20Document%20V%202%20Sept%2021%202021.pdf (accessed July 15, 2025).

COGR. 2022. *Research security and the cost of compliance phase I report.* https://www.cogr.edu/sites/default/files/Version%20Dec%205%202022%20research%20security%20costs%20survey%20FINAL.pdf (accessed July 11, 2025).

COGR. 2023. *Data management and sharing (DMS) and the cost of compliance.* https://www.cogr.edu/sites/default/files/DMS_Cost_of_Compl_May11_2023_FINAL%20%281%29.pdf (accessed July 11, 2025).

COGR. 2025. *Actionable ideas to improve government efficiency affecting the performance of research.* https://www.cogr.edu/sites/default/files/Actionable Ideas to Improve Gov Efficiency COGR_0.pdf (accessed June 12, 2025).

CRS (Congressional Research Service). 2017. *Federally funded academic research requirements: Background and issues in brief.* https://sgp.fas.org/crs/misc/R44774.pdf (accessed June 23, 2025).

Dorgelo, C., and J. Leibenluft. 2025. *DOGE interference in federal grantmaking adds burden, uncertainty, and risk.* Center on Budget and Policy Priorities. https://www.cbpp.org/sites/default/files/5-28-25bud.pdf (accessed August 19, 2025).

EAB (Education Advisory Board). 2025. Responding to federal policy shocks: What we learned from nearly 50 higher ed institutions. *Higher Education Strategy Blog*, June 10, 2025. https://eab.com/resources/blog/strategy-blog/responding-federal-policy-shocks-higher-ed-institutions/ (accessed August 19, 2025).

FDP (Federal Demonstration Partnership). 2014. 2012 faculty workload survey: Research report. https://thefdp.org/wp-content/uploads/fws_2012_final_rpt.pdf (accessed June 12, 2025).

FDP. 2020. *Faculty workload survey: Primary findings.* https://thefdp.org/wp-content/uploads/FDP-FWS-2018-Primary-Report.pdf (accessed June 12, 2025).

GAO (U.S. Government Accountability Office). 2016. *Federal research grants: Opportunities remain for agencies to streamline administrative requirements.* https://www.gao.gov/products/gao-16-573 (accessed June 12, 2025).

GAO. 2021. *Federal research grants: OMB should take steps to establish the Research Policy Board.* https://www.gao.gov/assets/gao-21-232r.pdf (accessed June 23, 2025).

Institute for Responsive Government. 2025. Reducing administrative burdens - the U.S. Federal government framework. https://digitalgovernmenthub.org/wp-content/uploads/2025/02/20250115-Reducing-Administrative-Burdens-6.pdf (accessed June 24, 2025).

Jager, B. 2023. Building a culture of compliance at liberal arts colleges and predominantly undergraduate institutions. *Journal of Research Administration*, 54(1), 66-86.

NASEM (National Academies of Sciences, Engineering, and Medicine). 2009a. *Ensuring the integrity, accessibility, and stewardship of research data in the digital age.* Washington, DC: The National Academies Press.

NASEM. 2009b. *Beyond 'Fortress America': National security controls on science and technology in a globalized world.* Washington, DC: The National Academies Press.

NASEM. 2016. *Optimizing the nation's investment in academic research: A new regulatory framework for the 21st century.* Washington, DC: The National Academies Press.

NASEM. 2022. *Protecting U.S. technological advantage.* Washington, DC: The National Academies Press.

NASEM. 2025. *Artificial intelligence and the future of work.* Washington, DC: The National Academies Press.

NCSES (National Center for Science and Engineering Statistics). 2021. *Higher education research and development: Fiscal year 2020.* https://ncses.nsf.gov/pubs/nsf22311 (accessed July 15, 2025).

NIH OLAW (National Institutes of Health Office of Laboratory Animal Welfare). 2024. *Animal Welfare Act and regulations (administered by USDA).* https://olaw.nih.gov/policies-laws/animal-welfare-act (accessed June 12, 2025).

NSF (National Science Foundation). 2024. *Trusted research using safeguards and transparency (TRUST)*. https://nsf-gov-resources.nsf.gov/files/NSF OCRSSP TRUST Policy Memo.pdf (accessed June 24, 2025).

NSF and NSB (National Science Board). 2014. *Reducing investigators' administrative workload for federally funded research*. https://nsf-gov-resources.nsf.gov/pubs/2014/nsb1418/nsb1418.pdf?VersionId=O0PSt6QhRdM4mTWI6j6jSTCEuD4EZghu (accessed June 24, 2025).

NSTC (National Science and Technology Council). 2022. *Guidance for implementing National Security Presidential Memorandum-33 (NSPM-33) on national security strategy for United States government supported research and development*. https://bidenwhitehouse.archives.gov/wp-content/uploads/2022/01/010422-NSPM-33-Implementation-Guidance.pdf (accessed August 5, 2025).

ORI (Office of Research Integrity). n.d. Conflicts of commitment. https://ori.hhs.gov/education/products/rcradmin/topics/coi/tutorial_4.shtml (accessed July 16, 2025).

ORI. 2024. *ORI final rule*. https://ori.hhs.gov/blog/ori-final-rule (accessed August 6, 2025).

OSTP (Office of Science and Technology Policy). 2025. *Agency guidance for implementing gold standard science in the conduct and management of scientific activities*. https://www.whitehouse.gov/wp-content/uploads/2025/03/OSTP-Guidance-for-GSS-June-2025.pdf (accessed July 15, 2025).

Rockwell, S. 2009. The FDP faculty burden survey. *Research Management Review* 16(2):29–44.

The White House. 2025a. *Remarks by Director Kratsios at the National Academy of Sciences*. https://www.whitehouse.gov/briefings-statements/2025/05/remarks-by-director-kratsios-at-the-national-academy-of-sciences/ (accessed June 12, 2025).

The White House. 2025b. *Fact sheet: President Donald J. Trump launches massive 10-to-1 deregulation initiative*. https://www.whitehouse.gov/fact-sheets/2025/01/fact-sheet-president-donald-j-trump-launches-massive-10-to-1-deregulation-initiative/ (accessed June 12, 2025).

2

Options to Optimize the Research Enterprise

Despite the challenges to achieve greater regulatory efficiency outlined in Chapter 1, the time and environment are right to modify current administrative and regulatory policies to better ensure that the research community is maximally productive while simultaneously maintaining the safety, accountability, security, and ethical conduct of publicly funded research. Because there are many routes to accomplish these goals, the committee is presenting a menu of options for consideration in each area. The options the committee presents in this report offer multiple potential solutions to each problem, and policymakers can choose the approach they determine is best suited to their needs. The committee also acknowledges that the options presented here come at a time when the federal workforce is downsizing. While these options aim to reduce workload, many do incur upfront costs in time and other resources. However, these upfront investments, if implemented with intention, would ultimately reduce administrative workload for researchers and hopefully for federal agencies as well.

This chapter is divided into topical areas for categories of research policies that need to be addressed, and adopting any of the options presented in each section would make research more productive. Each topical area in this chapter can be read independent of the others, and common themes arise in the approaches presented in the options across different topical areas. There are many options presented in this chapter, and the committee does not place them in any order of preference. Instead, those looking to implement solutions to the problems outlined should consider the right approach from the options presented. To inform decisions, the committee provides pros and cons for each option and potential timelines for implementation wherever applicable.

A NEW APPROACH TO RESEARCH REGULATIONS AND REQUIREMENTS

Historically, different federal agencies, each with its own mission, have supported a robust and world-leading U.S. scientific enterprise, creating a strong foundation of scientific knowledge and spurring technological innovation that has underpinned U.S. economic prosperity and national security. At the same time, this approach has led to a growing administrative burden, as Chapter 1 discussed. Solving the administrative challenges a researcher faces while navigating a decentralized U.S. research enterprise will require streamlining processes, establishing new coordinating functions, and regularly updating policies, requirements, guidance, and processes. Progress will require a new mindset from research funders and institutions, trust among the parties that everyone is acting in good faith toward a common good, and ongoing assessment of regulatory workload as changes are implemented.

Change should be made deliberately and thoughtfully, taking great care to not harm the U.S. research and innovation enterprise irreparably. Many of the options the report presents draw from those suggested by previous groups, and those groups have put significant and thoughtful analysis into them. At the same time, change is challenging to implement and has unintended consequences. It is important to acknowledge and address this whenever possible with utmost attention to preserving the strength of the U.S. research enterprise.

In developing the options for action, the committee found the same types of issues occurring in multiple oversight and regulatory areas. Although implementing any of the options within an area will lead to progress, three overarching principles should guide future decision-making. These principles, reflected in the 53 options presented in this chapter, represent a framework for developing additional options in the future.

1. **Harmonize regulations and requirements across federal and state agencies and research institutions.** The U.S. government should reduce administrative workload for the researchers it supports by harmonizing policies to the greatest extent possible across

all agencies. This may at times require compromising in the name of harmonization on the type, specificity, and format of information that a given agency requests. For example, a researcher should be able to prepare one standardized biographical statement for any federal grant proposal without deviation, with the same information required for all agency sponsors, and submit their application through one federal platform, such as the Science Experts Network Curriculum Vitae (SciENcv) system that the National Institutes of Health (NIH) hosts. When implementing a new regulation or requirement, it is important to consider whether any incremental gain in oversight that one agency might realize outweighs the cumulative costs to the U.S. research enterprise.

Federal requirements are not the sole source of increased burdens on researchers. Some states regulate research beyond what federal agencies require, which can create confusion and duplicative and additional requirements that may even conflict with federal requirements. Similarly, academic research institutions at times add unnecessary administrative work by implementing policies and procedures that go beyond the compliance requirements set by federal agencies. Along with necessary harmonization efforts at the federal level, states and institutions can also help reduce administrative burdens by harmonizing their policies and procedures to more closely adhere to federal requirements and not adopting additional compliance obligations. Coordinated community-based resources—such as those being developed by the research community via the U.S. National Science Foundation (NSF)-backed Safeguarding the Entire Community in the U.S. Research Ecosystem (SECURE) Center to address research security or the Compliance Unit Standard Procedures online repository for best practices in animal care and welfare—can help harmonize and enhance the quality of policies, procedures, guidance, and resources across the regulated community.

2. **Take an approach tiered to the risks involved when considering a new regulation or requirement.** Regulatory requirements need to be sufficient to ensure the safety of human research participants, the public, and those conducting the research; safeguard the fiscal integrity of federally funded research; ensure the appropriate care of research animals; and protect intellectual property and national security. At the same time, they should be calibrated to

the level of risk arising from noncompliance. Taking an approach to compliance that is tiered to risk—where the rigor of regulatory requirements aligns with the level of risk of an activity to society or regulatory objectives—can minimize the impact of federal requirements on researcher workload. State governments can also reduce regulatory requirements by taking a risk-tiered approach to research regulations and by being aware of U.S. government requirements.

Heavy workloads around research administration are largely a result of important efforts to manage risks, which include financial, reputational, and legal risks for institutions, safety and security issues, and ethics. While it is up to institutions to navigate risk management appropriately and ensure compliance with even low-risk work, agencies can help reduce administrative workload by allowing tolerances around smaller financial transactions and exempting research activities that present lower safety, security, or ethical risk from oversight, for example. Streamlining, rolling back, or reaching new compromises around consistent processes will likely mean that certain risks are managed differently or to a lesser degree than they are presently. However, processes that slow research and innovation unnecessarily are themselves a risk to the scientific enterprise and need to be reined in to better realize the benefits of the research enterprise.

In developing appropriate risk-tiering systems or models, government actors have multiple models to consider for the best approach, including those that already exist in the realm of scientific research. For example, Institutional Review Board (IRB) review of human subjects research is separated into three categories—exempt, expedited, and full review—with different requirements and levels of review based on potential risk to participants. The U.S. Environmental Protection Agency also relies on environmental risk assessments to understand and categorize human health and ecological risks of potential stressors through a scientific process that considers the amount of a stressor present, the exposure, and the effect (EPA, 2025). In addition, the government can look to examples in the financial industry and others that frequently model and tier risk and rely upon tools such as scorecards and decision trees to accomplish this (Kiritz et al.,

2019). Once tiered, consistent review is also needed to ensure no new information warrants a reassessment of risk.
3. **Use technology to simplify the process of complying with regulations and requirements to the greatest extent possible.** Available technology that leverages artificial intelligence (AI) and machine learning can automate and may simplify regulatory compliance processes. By automating repetitive tasks, using cross-agency databases, and providing real-time data analysis through platforms like NSF SECURE, technology can minimize the risk of human error and ensure that compliance processes are both thorough and efficient. To ensure these efforts do not create burden in an attempt to reduce it, it is imperative to standardize requirements and implement appropriate safeguards for the use of technology and to ensure these tools are used ethically and are accessible to all who need them.

These three overarching principles are important guides for future decision-making. They will help realize the goal of reducing the burden of complying with the regulations and requirements while ensuring the safety, integrity, and efficiency of the U.S. research enterprise. To implement these principles, the committee suggests closely reviewing the options presented in the next section (System-Wide Change), which would allow for continuous monitoring of regulations and requirements across federal agencies to sustain a research ecosystem that is not overly burdensome.

SYSTEM-WIDE CHANGE

The following sections present key problems, and for each problem, the committee provides options with varying pros and cons. This first section addresses issues that cut across multiple domains. However, the sections that follow focus on specific regulatory areas and detail the key problems for researcher workload in each domain.

Problem: Insufficiently centralized U.S. government oversight of the regulatory environment has led to too many overly complex, duplicative, and occasionally contradictory regulations, requirements, and reporting processes across federal agencies.

Many agencies fund scientific research in the United States, enabling the government to support a broad spectrum of mission-specific funding and security postures. At the same time, this multiagency approach has resulted in tremendous variability in systems, policies, and processes across agencies. While some of this variance can be mission specific, much of it results from a lack of intentionality and harmonization in the absence of coordinating activities. Furthermore, in efforts to address research security, agencies such as the U.S. Department of Energy (DOE) and U.S. Department of Defense (DOD) have skirted oversight by the Office of Management and Budget's (OMB's) Office of Information and Regulatory Affairs (OIRA) by issuing internal policies that imposed criteria for assessing security risks resulting from potential conflicts driven by foreign talent recruitment programs (COGR, 2019; Crowell & Moring LLP, 2023). While well intentioned, the lack of oversight through typical OIRA processes prevents community feedback and the opportunity to streamline and harmonize new requirements.

A few coordinating bodies do exist that can enable a greater degree of harmonization across agencies. One such body is the National Science and Technology Council (NSTC), established by Executive Order (EO) 12881 in November 1993.[1] The NSTC consists of cabinet members and agency heads and is led by the assistant to the president for science and technology at the president's discretion. One of NSTC's functions is to coordinate the science and technology policymaking process across agencies. According to the EO, the council may function through established or ad hoc committees, task forces, or interagency groups. In the area of research security, the NSTC interagency coordinating function led to the development of National Security Presidential Memorandum-33 (NSPM-33)[2] and the subsequent development of implementation guidance and common definitions, along with the Common Current and Pending Support and Biosketch Forms and Research Security Program requirements. Federal agencies have continued to coordinate on implementing these measures outside of the NSTC process, working from the foundation laid there.

---

[1] Exec. Order No. 12881, *Establishment of the National Science and Technology Council*, C.F.R., title 3 (1993): 2450–2451. https://www.govinfo.gov/content/pkg/WCPD-1993-11-29/pdf/WCPD-1993-11-29-Pg2450.pdf.

[2] National Security. Presidential Memorandum No. 33, *United States Government-Supported Research & Development National Security Policy* (Jan. 14, 2021). https://trumpwhitehouse.archives.gov/presidential-actions/presidential-memorandum-united-states-government-supported-research-development-national-security-policy/.

TABLE 2-1 Options to Address Insufficiently Centralized U.S. Government Oversight of the Regulatory Environment

Option 1: Establish a permanent function within the Office of Management and Budget (OMB) with the authority to coordinate cross-agency requirements

Goal:
To ensure strong leadership and strategic focus, establish a permanent career role in OMB that is charged with coordinating cross-agency requirements that affect federally funded academic research and has the authority to ensure agency coordination.

Approach:
OMB could create a permanent assistant director for institutional research coordination and community engagement position to collaborate with the Office of Information and Regulatory Affairs (OIRA) and the Office of Science and Technology Policy, and use the National Science and Technology Council to institute harmonization. This individual, a nonpolitical appointee, would serve as a resource that members of the research enterprise would turn to for assistance when inconsistencies arise or are anticipated between different funders or regulatory programs. The role would include overseeing the development of policies and requirements and how agencies implement them and identifying means to accelerate agency implementation when necessary.

Pros:
- Clearly identified central point with the authority of the White House to ensure coordination.
- Allows for harmonizing new areas of regulation to prevent future administrative expansion.
- Provides for expedited workload reduction without congressional action.

Cons:
- Can be successful only with voluntary participation of federal agencies and voluntary willingness to change existing requirements in the interest of harmonization and burden reduction.

Option 2: Appoint a Federal Research Policy Board

Goal:
Increase harmonization across federal agencies. The committee sees value in having a board dedicated to facilitating agency coordination and reducing federally imposed administrative work rather than the current widespread practice of having each agency develop its own approach to regulatory compliance.

continued

TABLE 2-1 Continued

Approach:
One key recommendation from the National Academies' 2016 report[a] was for Congress to establish a Research Policy Board composed of representatives from federal funding agencies and academic research institutions to "make recommendations concerning the conception, development, and harmonization of policies having similar purposes across research funding agencies." The year the report was released, Congress passed legislation through the 21st Century Cures Act[b] that required OMB to set up a Research Policy Board no more than a year after enactment. However, OMB never established the Research Policy Board, and congressional authority for the board ended on September 21, 2021.

Congress can reauthorize the creation of a Research Policy Board within OIRA, granting it similar composition and authorities to those outlined in the 21st Century Cures Act.

Pros:	**Cons:**
• Clearly identified central point to recommend federal requirements requiring harmonization.	• Requires congressional action, which will likely take time to implement.
• Allows for input from academic research institutions in the research compliance process.	• Even if authorized, implementation is not assured. Congress has previously required establishing the Research Policy Board, but OMB never followed through.
• Allows for harmonizing new areas of regulation to prevent future administrative bloat.	• Federal Advisory Committee Act requirements may dampen federal enthusiasm for establishing this advisory board.
	• Board members alone will not have the knowledge to address the wide range of topics that would arise.
	• While the board can make recommendations, it would not have the broad authority or participation to ensure agency coordination.

Option 3: Use the Federal Demonstration Partnership (FDP) to explore innovative ideas and practices through pilot programs

Goal:
Establish low-risk processes for testing innovative approaches to increase harmonization and use of approaches that are tiered to risk.

TABLE 2-1 Continued

Approach:
FDP is "an association of federal agencies, research policy organizations, and academic research institutions with administrative, faculty, and technical representation."[c] FDP has a track record of successfully implementing processes that improve regulatory efficiency and reduce administrative workload.[d] Its purpose is to streamline the administration of federally funded research grants and to do so through demonstration projects, where FDP identifies, tests, and advances new directions and practices. For example, guided by an interagency working group, FDP collaborated with the National Center for Biotechnology Information at the National Institutes of Health to build the Science Experts Network Curriculum Vitae (SciENcv) system to reduce the work required to develop biosketches for grant proposals.[e] As agencies develop new models and innovative approaches, they could work directly with FDP to test and refine these tools before formally launching them. This is currently taking place with a flexible risk-tiered approach to cybersecurity being developed cooperatively between federal agencies, research institutions, and faculty representatives via the FDP to meet federal research security program requirements. Piloted approaches to implementing new requirements can be evaluated and modified prior to broader implementation.

Pros:	Cons:
• Provides a low-risk pilot model to inform policy and process development and implementation. • Can facilitate rapid diffusion of processes and generate feedback for federal implementation.	• Smaller institutions may lack resources to participate in pilots, skewing results toward well-resourced universities even when intentional efforts are made. • If agencies reject processes developed by pilots, burden will persist. Collaborative development with agencies can reduce the possibility of this outcome.

[a] National Academies of Sciences, Engineering, and Medicine. 2016. *Optimizing the nation's investment in academic research: A new regulatory framework for the 21st century.* Washington, DC: The National Academies Press.

[b] *21st Century Cures Act,* Public Law 114-255, 130 Stat. 1033 (December 13, 2016).

[c] FDP (Federal Demonstration Partnership). n.d. *Who we are.* https://thefdp.org/ (accessed July 11, 2025).

[d] FDP. 2025. *Organization history.* https://thefdp.org/organization/history/#tab-id-2 (accessed July 11, 2025).

[e] NIH (National Institutes of Health). 2025. *SciENcv background.* https://www.ncbi.nlm.nih.gov/sciencv/background/?hss_channel=lcp-9398777 (accessed August 11, 2025).

Given the broad reach of the federal regulatory environment and the vast number of regulations that affect researchers,[3] the committee chose to focus on areas where regulatory reform could have the biggest effect on reducing the burden on the nation's research community. The following sections identify specific problem areas and offer options to address those areas along with the pros and cons of each option where appropriate.

REGULATORY AREA 1: GRANT PROPOSALS AND MANAGEMENT

Virtually every federal funding agency has its own set of requirements for submitting a research proposal. In addition, multiple grant submission portals exist, each requiring administrators and researchers to meet varying conditions, learn different systems, and keep current with agency-specific system requirements (COGR, 2025a). Then, once an investigator receives a research grant, each federal research agency requires different, often redundant reporting requirements for disclosing inventions throughout the lifetime of an award, filing progress reports, and detailing how funds were spent. Each agency also has its own closeout requirements at the end of the award lifecycle.

There has been some progress over time addressing these problems. For example, in response to the Uniform Guidance[4,5] issued by OMB in December 2013, federal funding agencies began efforts to harmonize Notices of Funding Opportunities (NOFOs) and address the fact that federal agencies had developed different acronyms and formats to make funding announcements, such as Funding Opportunity Announcements, Requests for Applications, and Parent Announcements, which made it difficult for institutions to track and apply for these opportunities. The switch to NOFOs has eliminated confusion and a source of additional workload for research universities who notify their faculty of funding opportunities. The Grant Reporting Efficiency and Agreements Transparency Act of

---

[3] As part of its information gathering, the committee published a request for information seeking input from the research community on regulations that could be improved. The feedback ranged from requirements affecting a large number of researchers (like grant application processes) to regulations affecting a relatively small number of researchers (such as regulating zebrafish larvae).

[4] Uniform Guidance refers to the set of federal regulations that govern the administration of federal grants and cooperative agreements.

[5] *Federal Financial Assistance,* 2 C.F.R. Part 200 (April 22, 2024).

2019[6] sought to streamline federal grant reporting, but according to an analysis by the U.S. Government Accountability Office, OMB and the U.S. Department of Health and Human Services (HHS) have only partially met a statutory requirement to establish government-wide data standards for information reported by grant recipients and have not met the statutory requirement to jointly issue guidance to all agencies directing them to apply the data standards (GAO, 2024).

As the Council on Governmental Relations (COGR) and other professional associations have suggested, a single whole-of-government application process can eliminate the need for a large amount of duplicative, burdensome reporting requirements for researchers. The federal government can improve upon the compliance capabilities of existing databases such as SAM.gov (the government system for contracting, grants, loans, and other financial assistance), Research.gov (proposal and grant system for NSF awards), and ORCID.org (used to link funding and publication outputs to researcher identities) to facilitate development of a single whole-of-government portal (COGR, 2025a).

Instead of compounding training requirements for U.S. researchers, agencies could develop a unified set of skills and competencies across researchers and institutions and with adaptability for field- or method-specific needs. The government can also leverage a just-in-time approach, requesting additional action regarding training at the time of award. In addition, the committee offers the following pre- and post-award process structures to yield a more seamless experience between the government and the researcher.

In the pre-award stage, changes could include

- introducing a two-stage process to reduce time and effort for the initial submission;
- establishing a single portal as recommended by COGR; or
- reorganizing or reviewing the requirements to remove redundancy, prioritize consistency, and reduce researcher administrative workload (NASEM, 2016).

Currently, the process for reviewing and approving each agency's application form and progress report form falls under the Paperwork Reduction

---

[6] *Grant Reporting Efficiency and Agreements Transparency Act of 2019*, H.R. 150, 116th Congress (June 6, 2019).

Act,[7] which was amended most recently in 1995 to reduce the burden of government information collections on the American public. Under the act and its implementing regulations,[8] agencies must submit any collection of information that applies to 10 or more individuals for public comment and for approval by OIRA. For any continuing collection, such as with grant applications and progress reports, agencies must renew this approval every 3 years, and at each renewal the agency must solicit public comment and submit the application to OIRA for approval.

The 2016 National Academies of Sciences, Engineering, and Medicine report noted that agency submissions to OIRA for grant applications and other information collections associated with required recordkeeping for grants were inconsistent and varied widely by federal agency, potentially underestimating the burden on universities to comply with different required submissions (NASEM, 2016). OIRA could use the authority granted to it through the Paperwork Reduction Act to encourage federal agencies to centralize and harmonize grant application and reporting requirements.

The committee provides options for a two-stage award process, harmonization across agencies, reduction of duplicative efforts, updated requirements, and clearer notifications of policy changes to address these issues.

Problem: Inconsistent, complex, and burdensome grant processes across federal agencies.

Current grant application processes differ widely across funding agencies, with agencies requesting information that varies in specificity, timelines, and length in initial grant proposals. While systems such as Grants.gov, for example, may provide a clearinghouse for funding announcements across the federal government, allowing for a single portal for submission, notice of policy changes, and submission of reporting requirements, other grant management systems require researchers and administrative officers to interface with platforms that are often disconnected, increasing time spent on submission and compliance. Onerous and disjointed research grant processes have increased the time individual researchers and institutional pre- and post-award staff must spend on administrative actions to satisfy

---

[7] *Paperwork Reduction Act of 1980*, H.R. 6410, 96th Congress (December 11, 1980).
[8] *Controlling Paperwork Burdens on the Public,* 5 C.F.R. Part 320 (August 29, 2005).

TABLE 2-2 Options to Address the Burdensome Grant Processes (Regulatory Area 1)

Option 1.1: Introduce a federal-wide, two-stage pre-award process

Goal:
This option would improve the efficiency of the pre-award process and harmonize the grant application process across funding agencies.

Approach:
To reduce the time spent by researchers on proposal preparation and submission, agencies could operationalize a common "letter of intent" (LOI) mechanism across funding opportunities that applies to single and multisite research, centers, instrumentation, training, and other grant mechanisms deemed appropriate by agency officials. LOIs could consist of five- to seven-page proposals focused on scientific merit, similar to the LOIs the U.S. National Science Foundation (NSF) uses. Subject-specific review panels would evaluate the LOIs to assess the merit of the proposed research. To ensure consistency across federal agencies, the Office of Information and Regulatory Affairs (OIRA) could review LOI proposal templates to reduce redundancies and harmonize formats, potentially through collaborative work with the Federal Demonstration Partnership. Researchers who submitted LOIs deemed high quality per the recommendation of the subject-specific review panel and identified as likely to be recommended for funding would then be invited to submit full-length proposals with detailed budget justifications and additional scientific detail, publication history, and other required documents to satisfy statutory and legal requirements. The expanded proposals would then undergo a final review to identify the most meritorious projects for final funding decisions.

Pros:
- Reduces administrative effort and time spent on proposals, the majority of which will not be funded given limited resources.
- Provides agency program officers and reviewers with clarity on the scientific merit of potential proposals at the beginning of the decision-making process and provides feedback to improve proposal quality.
- Reduces the time and effort for reviewers assessing the viability of proposed research.
- Reduces the time spent developing research proposals for researchers and institutions.

Cons:
- Potential increases in LOI rejection appeals and adjudications.
- Potential increases in requests for more extensive review in the grant proposal process, further increasing the time to award.
- This has been proposed widely but has not been adopted by all agencies, suggesting potential barriers that would need to be addressed.

continued

TABLE 2-2 Continued

Option 1.2: Harmonize application and reporting systems including leveraging artificial intelligence (AI)–enabled tools

Goal:
Develop a single platform to minimize redundancy and leverage innovation to automate information gathering for reporting and generate common information needs for funding agencies.

Approach:
All funding agencies could leverage and expand their use of existing platforms such as Research.gov, which is used by NSF; the Open Researcher and Contributor ID; and Science Experts Network Curriculum Vitae (SciENcv) system. Federal agencies could also use standard reporting formats consistent with the requested information on research goals, outcomes, societal benefits, and resources used. Developing and implementing machine-readable form standards across federal agencies, providing access to AI tools to institutions, and integrating AI into grant applications and reporting systems, with appropriate safeguards and policies, would enable AI to parse and automate grant materials.

Pros:
- Supports comparability across agencies and improves transparency and outcome tracking.
- Reduces time spent on duplicative forms and information requests.
- Supports the evaluation of federal research programs and outcomes.

Cons:
- Continues to rely on human review of AI for accuracy.
- Impacts from agencies that may request exemptions to harmonized processes.
- Raises concerns of over-centralization and data security on intellectual property and individual data.
- High cost to develop an appropriate AI model.

Option 1.3: Prioritize the reduction of duplication and unnecessary burden in the research enterprise in OIRA review of research agency information collections under the Paperwork Reduction Act[a]

Goal:
Use existing authority to require harmonization of grant application and reporting requirements across federal agencies.

TABLE 2-2 Continued

Approach:
Given that agencies must resubmit their information collections to OIRA every 3 years, and that when submitted, the Office of Management and Budget (OMB) has a 30-day public comment period on the information collection, entities affected by these collections could submit public comments to the respective agency conducting information collections and OIRA about agency burden estimates and duplications across agencies. In response to these comments, OIRA could require that agencies respond before approval. OIRA could also condition approval of reporting and recordkeeping requirements on agency changes to the forms that would reduce burden and eliminate duplication.

Pros:	Cons:
• There would be no need for statutory or regulatory changes. The Paperwork Reduction Act is an existing law, and all that would be required is revised enforcement of it by OIRA. • OIRA also has regulatory review authority, and the current administration's emphasis on eliminating regulations gives OIRA an opportunity to push agencies to review their existing corpus of requirements in conjunction with Paperwork Reduction Act review. • The Paperwork Reduction Act has the built-in public participation requirement noted above, which provides a mechanism for those affected by these requirements to highlight their flaws.	• OIRA appears to be understaffed for certain purposes, so its Paperwork Reduction Act reviews, which may fall below regulatory review and other statutory obligations, can add considerable delay.[b] • OIRA review under the Paperwork Reduction Act is also particularly reliant on public comment. If research institutions do not highlight the problems with research-related information collection and recordkeeping requirements, OIRA is unlikely to unearth them on its own. • The Paperwork Reduction Act process is burdensome for agencies. It can take 6 to 9 months for an agency to secure approval of an information collection.[c] As a result, the process often becomes a box-checking enterprise to which neither OIRA nor agencies pay requisite attention.

Option 1.4: Update cost accounting and financial compliance requirements

Goal:
Modernize existing overly complicated cost accounting and compliance requirements while maintaining adequate financial reporting.

continued

TABLE 2-2 Continued

Approach:
The 2016 report recommended that OMB update and amend the Uniform Guidance,[d] in consultation with research institutions, to eliminate institutional expectation for filing financial disclosure statements at every transaction in favor of periodic updates on a pre-determined schedule. Federal audit agencies could give due consideration to pre-existing commercial and university recordkeeping systems and generally accepted accounting principles to determine whether they can provide the government with the assurance it requires. In addition, OMB could propose a centralized disclosure system that satisfies financial reporting requirements across multiple grants and agencies.

Pros:
- Decreases administrative workload for institutional financial offices.
- Harmonizes compliance directives across grants and agencies.

Cons:
- May affect the ability of agencies to detect financial inconsistencies.
- Requires integration with audit systems.

Option 1.5: Update SAM.gov with notifications when policy changes are made that impact award requirements

Goal:
Reduce confusion for grant managers and investigators and increase timely compliance with federal grant requirements.

Approach:
As policies and regulations change, federal agencies update requirements for compliance in Grants.gov. However, it is difficult for grant managers and investigators to keep up with grant management requirements and their changes across federal agencies. To streamline registration and renewal processes, the General Services Administration could update the login and notification system in SAM.gov to facilitate a more user-friendly interface for major updates on policies and how they affect awards on an annual basis.

Pros:
- Decreases workload on grant managers and investigators to keep up with policy changes.
- Increases timely compliance with grant management requirements.

Cons:
- Requires resources to regularly provide updates.

[a] *Paperwork Reduction Act of 1980*, H.R. 6410, 96th Congress (December 11, 1980).

[b] Rudalevige, A. 2018. Regulation beyond structure and process. *National Affairs*. https://www.nationalaffairs.com/publications/detail/regulation-beyond-structure-and-process (accessed July 2, 2025).

[c] ACUS (Administrative Conference of the United States). n.d. *Improving the efficiency of the Paperwork Reduction Act*. https://www.nationalaffairs.com/publications/detail/regulation-beyond-structure-and-process (accessed July 2, 2025).

[d] *Uniform Administrative Requirements, Cost Principles, and Audit Requirements for Federal Awards*, 89 Fed. Reg. 99695 (December 12, 2024).

often redundant requirements across funding sources, diverting time and resources from pursuing additional funding and conducting research. Researchers and institutions have to spend unnecessary resources and time addressing requirements that do not directly improve research quality or integrity. Lengthy grant submissions are especially challenging for researchers in an environment where success rates are low. In 2024, only 20 percent of NIH proposals were funded (NIH OER, 2024) and 26 percent of NSF proposals (NSF, 2025).

REGULATORY AREA 2: RESEARCH MISCONDUCT

The 2000 revision of the Federal Research Misconduct Policy[9] establishes that federal agencies have authority over research misconduct, while research institutions bear the responsibility for preventing and detecting it. This policy defines research misconduct as falsification, fabrication, and plagiarism in proposing, performing, or reviewing research, or in reporting research results. Such oversight is vital to ensuring accurate and ethical conduct of federally funded research and maintaining trust with the public, policymakers, and the broader scientific community.

The committee found, however, that current regulations create challenges that do not further these important goals as effectively or efficiently as they could. When the Federal Research Misconduct Policy was revised in December 2000, many comments were received about ensuring uniform implementation of the policy across agencies (ORI, 2000). The reality of implementation, however, has not resulted in uniformity.

While some agencies rely entirely on the Federal Research Misconduct Policy, others have created their own detailed procedures (VA and VHA, 2017). The U.S. Department of Veterans Affairs (VA) Directive 1058.02, for example, has become increasingly detailed over time, losing much of the flexibility the Federal Research Misconduct Policy incorporates (Defino, 2025).

Agency variation can also result in inconsistency and differing standards for responding to misconduct. To modernize the existing research misconduct rule and provide better oversight, the HHS Office of Research Integrity (ORI) issued a "Final Rule" in 2024 that updated regulations established in 2005 (42 CFR Part 93; ORI, 2024). This update modern-

---

[9] Executive Office of the President, Office of Science and Technology Policy, *Federal Policy on Research Misconduct; Preamble for Research Misconduct Policy*, 65 Fed. Reg. 76260 (October 14, 1999).

ized definitions and procedures related to plagiarism and fabrication and falsification of data, streamlined certain parts of the investigatory process, and introduced new timelines and record-keeping standards. The Final Rule also aims to increase transparency by tasking institutions with maintaining detailed records of incidents of misconduct, submitting compliance reports by 2025, and prioritizing training and institutional accountability for ensuring reproducibility. ORI's Final Rule, however, currently pertains only to HHS grantees and does not align with the corresponding rules of other federal funding agencies such as NSF, the U.S. Department of Agriculture (USDA), and others, which either use their own research misconduct regulations and frameworks or rely on the Federal Research Misconduct Policy released in 2000.[10]

Many researchers are funded by multiple agencies and therefore must adhere to their different requirements. The variation in misconduct policies and procedures, such as investigation timelines, the length of reviews by agencies such as ORI, and the increased burden on institutions to retain more information, makes it challenging for researchers, especially when an allegation of misconduct is received involving multiple funding agencies.

In such instances, agencies have to agree on which agency is the "lead" agency; otherwise, each involved agency may have to evaluate the allegation separately. The same is true of overlapping inspectors general, audit agencies, and investigative agencies that may compete for a role in the enforcement process, creating further variation and potential inconsistency across similar misconduct cases where a different set of agencies are involved. In addition, to manage funding agencies' variations, institutions often adopt the strictest standard for responding to allegations of research misconduct, losing flexibility when a less strict but more appropriate standard could apply beyond research misconduct. HHS has recently introduced new guidelines to improve efficiency, but the effect of these is not yet known.

The committee also noted challenges in current reporting structures. At present, much of the infrastructure for managing and addressing research misconduct cases relies on email rather than secure and standardized online portals. This can lead to delays, lost communications, leaks, and issues with attorney-client privilege. In certain instances, submissions require specific formatting, which takes time to generate in personal emails; this format could be applied automatically via an online standardized form.

---

[10] Executive Office of the President, Office of Science and Technology Policy, *Federal Policy on Research Misconduct; Preamble for Research Misconduct Policy*, 65 Fed. Reg. 76260 (October 14, 1999).

The committee has heard of cases dragging on for months and even years as communication is bogged down by inefficient systems.

The committee provides options to address these issues through harmonization, assessment of new guidance, new systems, and more efficient approaches to managing misconduct cases.

Problem: Different standards for research misconduct proceedings across agencies.

While a single Federal Research Misconduct Policy exists, agency variation continues, with particularly inflexible requirements from HHS and VA Directive 1058.02.[11] This agency variation creates a number of problems, including potential inconsistencies in how misconduct cases are handled and a lack of flexibility to apply the most appropriate standard to the case at hand. There are increased challenges when multiple agencies are involved and need to determine a lead agency to oversee a case.

TABLE 2-3 Options to Address the Problem of Different Standards for Research Misconduct Proceedings Across Agencies (Regulatory Area 2)

Option 2.1: Agencies follow a single flexible federal misconduct policy

Goal:
Lower the administrative burden on institutions by simplifying and harmonizing research misconduct procedures across the federal funding agencies. Punitive standards could be applied consistently so that accused researchers are held to the same consequences.

Approach:
The Federal Research Misconduct Policy could be revised to provide a single set of flexible requirements to be followed for handling research misconduct allegations rather than vesting authority for research misconduct in each federal agency. Each federal funding agency can defer to the Federal Research Misconduct Policy and eliminate their separate procedures.

continued

---

[11] *Research Misconduct,* Veterans Health Administration Handbook 1058.02 (February 7, 2014).

TABLE 2-3 Continued

Pros:	Cons:
• Streamlines response to research misconduct allegations by ensuring consistency across federal agencies. • Increases a shared understanding of research misconduct proceedings. • Allows greater flexibility for both agencies and institutions.	• Requires regulatory, policy, and procedural changes across multiple agencies. • Requires agreement across agencies for a common framework instead of agency-specific variation.

Option 2.2: Ensure a single lead agency has jurisdiction over misconduct allegations for research with multiple funding agencies

Goal:
Increase clarity of which agency's procedures apply for a given misconduct case and reduce duplicative oversight.

Approach:
Current Federal Research Misconduct Policy states that "if more than one federal agency has jurisdiction over allegations of research misconduct, those agencies could work together to designate a lead agency."[a] While this intends to ensure a single agency has oversight over a misconduct case involving research funded by multiple agencies, it is not a requirement. The policy could be changed to require a lead agency and set clear criteria for how a lead agency would be determined.

Pros:	Cons:
• Reduces uncertainty about which agency guidelines to follow in misconduct cases. • Prevents potentially having to deal with multiple agency guidelines.	• Still allows agency variation in approaches to misconduct. • Does not prevent differing standards being applied to misconduct allegations based on agency guidelines and policies.

[a] ORI (The Office of Research Integrity). 2000. *Federal research misconduct policy.* https://ori.hhs.gov/federal-research-misconduct-policy (accessed July 11, 2025).

Problem: Slow and ineffective digital infrastructure for handling misconduct cases.

Current communications with federal funding agencies for handling research misconduct proceedings rely heavily on an email system plagued by slow response times and communication issues. This introduces challenges for institutions and agencies tracking any single case and increases the time between reporting of research misconduct case decisions by institutions to determinations and corrective actions from the funding agency. In addition, certain reports, such as the separate inquiry and investigation reports for HHS, require incredibly specific formatting that is left to institutions to implement.[12] These challenges are something that ORI is aware of and working actively to improve.

---

[12] For more information on the requirements of the HHS inquiry report, see 42 C.F.R. § 93.307(d).

TABLE 2-4 Options to Address the Problem of Slow and Ineffective Digital Infrastructures for Handling Misconduct Cases (Regulatory Area 2)

Option 2.3: Develop a new system for managing misconduct cases

Goal:
Improve infrastructure for reporting and managing research misconduct proceedings to ensure these proceedings move more effectively and efficiently.

Approach:
Rather than relying on email communication to share key reports and information during misconduct proceedings, an existing or newly built system could be used to ensure the organization of all key files. This system could also ensure consistent formatting of all reports by providing online forms that institutions can fill out and will automatically be structured as the agency desires. In the cases of a harmonized process, this could be a single government-wide system. If the current multiagency system remains, each agency could create its own system.

Pros:	Cons:
• Ensures consistency in submitted forms to aid agency employees reviewing these forms while reducing time institutions need to spend on formatting.	• Without a single harmonized misconduct process across the U.S. government, multiple systems or approaches will be required to manage cases for different agencies.
• Improves organization of forms and reduces likelihood of forms and reports getting lost.	• Requires time, resources, and expertise to build a new system for reporting and managing misconduct proceedings that allows for case-specific, tailored information and for thorough investigations.

Option 2.4: Continue efforts to improve response times

Goal:
Reduce the amount of time spent waiting for response to move misconduct cases forward.

Approach:
The U.S. Department of Health and Human Services' Office of Research Integrity is already working to reduce the amount of time for responses to emails regarding research misconduct cases. These efforts could continue and increase with a clear target set for a timeline for accurate response. No new system would be built, but the old approach would be maintained with efforts to improve efficiency.

Pros:	Cons:
• Does not require time and resources to change the current structure for managing misconduct proceedings.	• Maintains current system which does not address formatting inefficiencies and inconsistencies.
• Shortens proceedings timelines.	

Problem: Uncertain impact of new HHS requirements.

HHS recently introduced new regulations to address and respond to research misconduct (ORI, 2025) in an effort to improve efficiency and streamline processes. These regulations will not be in full effect until 2026, so there is no available assessment of their effectiveness. Early reviews, however, suggest that while some updates represent welcome change and streamlining, other changes may not be consistent with improving efficiency (Green et al., 2023; Phillips and Earl, 2025). Some mechanism is needed to ensure these changes are in fact having the desired effect.

In addition, HHS has strict procedures for research misconduct, and because of the amount of funding for research from HHS, most institutions adopt its framework. Given this, even without harmonizing regulations across agencies, evaluating and streamlining HHS requirements would help reduce workload and burden.

TABLE 2-5 Option to Address the Uncertain Impact of New HHS Guidelines (Regulatory Area 2)

Option 2.5: Assess new U.S. Department of Health and Human Services (HHS) misconduct regulations for efficiency and effectiveness
Goal: Determine the effectiveness of new HHS guidelines and the impact of efforts to improve efficiency in the current HHS structure.
Approach: The HHS Office of Research Integrity could review the efficiency and effectiveness of the new regulations at a predetermined future time, ideally within 5 years of full implementation, and make adjustments based on feedback from key stakeholders.

Pros:	Cons:
• Creates predetermined timeline for assessment to ensure changes are having the desired effects.	• Requires agency time and resources for review. • In the absence of efforts at harmonization, does not reduce duplicative and inconsistent agency approaches to misconduct.

REGULATORY AREA 3: FINANCIAL CONFLICT OF INTEREST IN RESEARCH

As with research misconduct policies, there is not a harmonized financial conflict of interest (FCOI) in research policy applied across funding agencies. FCOI is defined as a significant financial interest (SFI) that affects the design, conduct, or reporting of federally funded research (NIH, 2024a). Federal policies often provide monetary thresholds for what they determine to be an SFI that may affect a given research project. In response to a series of scientific misconduct and research harm cases in the 1970s and 1980s, Congress directed HHS to develop regulations addressing FCOIs. The Public Health Service (PHS) released the first FCOI regulation in 1995, with HHS revising it in 2011.

While most institutions have standardized their internal FCOI policies for awards to align with PHS requirements, each federal agency has developed its own FCOI policy and procedures, disclosure thresholds, and reporting requirements. In addition, the policies often conflict with each other, requiring institutions to create manual systems or use the most stringent requirements for disclosure, adding additional work for researchers and reviewers (COGR, 2025a). NSF's policy, which it adopted from the original 1995 PHS policy, has a narrower definition of a reportable FCOI than the current PHS policy, is less burdensome than other reporting requirements, and has a different threshold than, for example, the U.S. Food and Drug Administration (FDA). The PHS threshold is $5,000, while the FDA threshold is $25,000, and neither are adjusted for inflation over time.

A survey by the Association of American Medical Colleges found that when PHS lowered its definition of an SFI from $10,000 to $5,000, 49 percent of the responding institutions reported an increase in the number of SFI disclosures (AAMC, 2020). However, there was only a 13 percent increase in the number of FCOIs reported to PHS and a decrease in the percentage of SFIs found to be FCOIs. This suggests that lowering the SFI threshold, with no indexing to inflation, may not be meeting the intended goal of protecting research integrity and that agencies could use an FCOI model with a higher SFI threshold, such as the NSF model (COGR, 2025a). Moreover, requiring actions such as conducting exhaustive retrospective reviews for determining bias, or reporting FCOIs through the eRA Commons platform, the online interface for NIH grant applicants, has not significantly addressed government concerns, with the former rarely identifying bias but increasing workload significantly and the latter resulting in no action from the receiving agency (AAMC, 2020). The committee's options, outlined below, focus on creating uniform policies, adjusting FCOI thresholds, and eliminating unnecessary reporting and requirements.

Problem: Inconsistent FCOI in research procedures and requirements.

Five federal agencies maintain FCOI regulations or policies that are inconsistent in terms of financial threshold for disclosure: PHS and DOE (DOE, 2021), which have a threshold of $5,000;[13] FDA, which has a threshold of $25,000 (FDA, 2013); and NASA and NSF, which have a threshold of $10,000 (NASA, 2023; NSF, n.d.-a). The five agencies also have different reporting requirements. For example, PHS requires reporting of some FCOIs, while FDA requires reporting to the study sponsor. It is unclear what PHS does with these reported FCOIs, as agency guidance states only that program officers may request more information on a reported FCOI, but there is no guidance as to what might prompt such an inquiry (eRA, 2025). Multiple disclosure thresholds and reporting requirements, from these agencies as well as others with separate policies, increase the burden on researchers and administrative staff.

---

[13] *Promoting Objectivity in Research*, 42 C.F.R. 50 Subpart F (August 25, 2011).

TABLE 2-6 Options to Address the Inconsistent FCOI in Research Procedures (Regulatory Area 3)

Option 3.1: Create a uniform financial conflict of interest (FCOI) in research policy

Goal:
Harmonize existing policies and procedures across federal funding agencies to provide for the consistent application of FCOI disclosures, identification, and management regardless of funding source.

Approach:
Federal funding agencies can be directed to adopt one standard for FCOI in research, preferably the U.S. National Science Foundation (NSF) policy. Given NSF's broad reach across disciplinary areas, adoption of NSF's FCOI policies will ensure interagency alignment under a proven framework that maintains research integrity under a harmonized approach.

Pros:
- Provides for the consistent application of FCOI disclosures, identification, and management regardless of the federal research funding source.
- Allows institutions to simplify and harmonize current internal policy requirements.
- Ensures researchers are subject to one set of federal FCOI requirements if they move between institutions and/or participate in different research projects.

Cons:
- Requires buy-in from federal agencies to ensure adoption.
- Significant work was previously conducted to achieve this via an interagency group led by the Office of Science and Technology Policy, including assessment of current agency requirements and the development of a common policy and terms, but it was unsuccessful.

Option 3.2: Revert to the earlier $10,000 Public Health Service (PHS) significant financial interest (SFI) threshold with periodic adjustments for inflation

Goal:
Increase the reporting threshold for financial conflicts to $10,000, as it was prior to 2011,[a] with annual adjustments made for inflation similar to the threshold for reporting in the Centers for Medicare & Medicaid Services Open Payments System.

Approach:
PHS can increase the threshold for an SFI to $10,000 and publish an annual updated threshold amount, through the notice of opportunity to transition process, to account for inflation.[b] Alternately, PHS could maintain a lower SFI threshold but would update it periodically to account for inflation.[a]

TABLE 2-6 Continued

Pros:	Cons:
• Redirects focus on more highly compensated financial interests. • Future-proofs threshold in a systematic way by incorporating inflationary adjustments consistently across agencies.	• Requires PHS to establish a new, albeit simple, procedure for adjusting the SFI threshold for inflation. • PHS has argued against increasing the threshold in the past, citing differences in the "nature, scope, and applicability of Federal disclosure requirements."[c]

Option 3.3: Eliminate reporting of FCOIs for PHS-funded research

Goal:
Eliminate the reporting of FCOIs identified by academic research institutions to PHS.

Approach:
Institutions are required to identify and manage FCOIs and report noncompliance to federal agencies, per regulation.[a] However, PHS goes beyond the regulation to require reporting of FCOIs. As PHS is the only federal agency to require reporting of FCOIs and it is unclear what is done with these reports, PHS can remove the requirement to report elements of management plans for PHS-supported researchers.[a]

Pros:	Cons:
• Eliminates the additional work needed to revise existing documents for reporting purposes. • Eliminates the duplication of information received by PHS through the FCOI reporting and Pending and Other Support processes, which are not mirrored by other agencies.	• May be criticized for the optics of fewer FCOI reporting requirements.

Option 3.4: Eliminate the requirement for a "determination of bias"

Goal:
Eliminate redundant requirements.

Approach:
PHS can eliminate the requirements around conducting a "determination of bias."[a] As noted in the 2020 Association of American Medical Colleges report, out of more than 100 noncompliance reviews, it is extremely rare for an institution to find "bias."[d]

continued

TABLE 2-6 Continued

Pros:	Cons:
• Eliminates burden placed on institutions, researchers, and program officers to identify and review results of "determinations of bias," consistent with published data that indicate that such determinations are of no added value.	

a Promoting Objectivity in Research, 42 C.F.R. 50 Subpart F (August 25, 2011).
b Can be done by revising 42 C.F.R. Part 50, Subpart F.
c Responsibility of Applicants for Promoting Objectivity in Research for Which Public Health Service Funding Is Sought and Responsible Prospective Contractors, Fed. Reg. 76, no. 165 (August 25, 2011).
d AAMC (Association of American Medical Colleges). 2020. *AAMC Conflicts of Interest Metrics Project - Measuring the impact of the Public Health Service regulations on conflicts of interest.* https://www.aamc.org/media/50386/download (accessed July 2, 2025).

REGULATORY AREA 4: PROTECTING RESEARCH ASSETS

The United States has a vested interest in regulating research to ensure the protection of research assets. The policy options to improve the requirements for the protection of research assets are grouped under Research Security, Export Controls, and Cybersecurity and Data Management. The committee details options that implement federal government-wide consistent processes, revise legal limitations on training, update key definitions, adopt risk-based approaches, and renew previous initiatives intended to streamline key processes.

Research Security

Growing concerns about research security and foreign interference entered broad public discussion in 2018. This, in conjunction with the importance of research security to the full breadth of academic research, led Congress and the Executive Branch to take a number of actions, including the development of NSPM-33[14] and the Creating Helpful Incentives to

---

[14] National Security Presidential Memorandum-33 (NSPM-33), *United States Government-Supported Research & Development National Security Policy* (January 14, 2021). https://trumpwhitehouse.archives.gov/presidential-actions/presidential-memorandum-united-states-government-supported-research-development-national-security-policy/.

Produce Semiconductors for America (CHIPS) and Science Act of 2022.[15] NSPM-33, issued in January 2021, instituted broad requirements for disclosure and research security infrastructure for recipients of federal research and development funds that exceed $50 million annually.[16] NSPM-33 directed federal agencies to create common forms for the disclosure of foreign affiliations, appointments, and funding sources and research organizations receiving over $50 million per year in federal research funding to implement a research security program that includes cybersecurity, foreign travel security, research security, and export control training (NSTC, 2022). The 2022 CHIPS and Science Act added additional mandates for training for personnel on federal awards and policy harmonization across agencies.

Despite efforts on the part of federal agencies and strong engagement with the research community, the U.S. research security landscape remains fragmented, requiring universities to reconcile overlapping or conflicting rules from multiple agencies. Researchers continue to confront inconsistent disclosure forms and systems—for example, often re-entering data already housed in SciENcv because agencies have not uniformly adopted the agreed upon "common forms" and SciENcv system. Another example is with "conflicts of commitment," a new concept for institutions when first introduced (COGR, 2021). Conflicts of commitment occur when a researcher dedicates time to personal activities in excess of institutional policy or that may detract from their professional responsibilities (ORI, n.d.). This has led some institutions to develop and implement conflict of commitment programs. The ambiguity of some of the requirements has led to variation in how institutions are developing infrastructure to comply with the requirements and how internal policies affect researchers and trainees at different institutions. Foreign gift and contract reporting is also duplicative across federal agencies. Institutions also lack an authoritative, consolidated hub for information, communication, resources, FAQs, and training, requiring them to track and piece together guidance, a particular challenge for smaller institutions and research programs. The nascent NSF-funded SECURE Center, authorized in the CHIPS and Science Act of 2022, is poised to address some of these issues by working with the community to design and develop resources to meet their research security needs and providing "shared tools, best practices, training, analyses and

---

[15] *CHIPS and Science Act*, Division B, Title VI, Subtitle D (Pub. L. 117-167, §§ 10631-10638). 42 U.S.C. §§ 19231-19237.

[16] All institutions that receive federal funding, regardless of funding threshold, are subject to certain aspects of NSPM-33, such as disclosure requirements.

other information, all delivered through a shared virtual environment" (NSF, 2024). The SECURE Center also has a mechanism for federal agency engagement through its U.S. Government (USG) Steering Committee, which is "composed of USG leaders and research security subject matter experts and chaired by NSF" (NSF, 2023). In tandem, NSF SECURE Analytics[17] is designed to provide data-driven insight into actions needed for risk mitigation, decision-making, protecting integrity, and facilitating collaborations between institutions.

Agencies implementing the CHIPS and Science Act training requirements currently require research security training requirements at the proposal stage, consistent with the statutory language, requiring even unfunded applicants to spend time addressing those requirements. Furthermore, reviews to identify research security risks associated with fundamental research proposals and requests for further clarification and mitigation are conducted with divergent risk-review criteria and mitigation rubrics across and within agencies, adding further unpredictability and confusion. This environment increases administrative workload, creates inefficiencies in the research process, and diverts resources from discovery and innovation. Coordinated federal leadership, harmonized agency implementation of research security requirements, just-in-time training, and common foundational principles for risk reviews that incorporate a tiered-risk approach to evaluations are needed to align protections with actual risk while minimizing unnecessary burden.

---

[17] For more information, see https://secure-analytics.org/.

TABLE 2-7 Options to Address Research Security Compliance Issues (Regulatory Area 4)

Option 4.1: Implement the National Security Presidential Memorandum-33 (NSPM-33) common disclosure forms and disclosure table without deviation as the primary means to identify and address conflicts of commitment (COCs) and develop federal-wide FAQs via the interagency working group; in addition, use the Science Experts Network Curriculum Vitae (SciENcv) system, persistent identifiers (PIDs), and application programming interfaces (APIs) across research funding agencies

Goal:
Reduce duplication and deviation in common disclosure forms by adopting the interagency developed "common" forms and utilizing SciENcv and PIDs to enable prepopulation of forms, thereby further and significantly reducing researchers' administrative workload.

Approach:
Agencies are still in the process of implementing the common federal-wide current and pending support and biosketch disclosure forms, published in November 2023.[a] Full implementation and avoiding agency-specific deviations will reduce administrative workloads for researchers. Deviations are taking the form of separate, additional requirements outside of the common forms. Although several agencies have indicated they will not adopt SciENcv at this time, use of this system will significantly reduce researchers' administrative workload and should be prioritized.

Federal-wide use of the NSPM-33 pre- and post-award disclosure table, housed on the U.S. National Science Foundation (NSF) website,[a] will reduce the potential for differing agency interpretations that lead to stakeholder confusion and increased workloads. So, too, will the development of federal-wide, rather than independent agency, disclosure FAQs. Agency use and requirement of PIDs, as outlined in NSPM-33, to prepopulate forms with relevant information, as well as APIs to capture information from institutions and other sources, will further reduce workload. To avoid confusion and minimize administrative workload, the common forms could be uniformly used for identifying COCs that should not be addressed in conflict-of-interest (COI) policy, such as through the U.S. Department of Energy's (DOE's) draft COI/COC policy.

Pros:	Cons:
• Using these standardized templates, guidance, and mechanisms for pre-population will reduce federal variation and simplify reporting. • Consistent requirements and ease of use will facilitate compliance.	• May face challenges by some agencies in implementing the common forms via SciENcv.

continued

TABLE 2-7 Continued

Option 4.2: Use the Safeguarding the Entire Community in the U.S. Research Ecosystem (SECURE) Center as an interactive research security information hub to keep the community current on the latest information and provide resources to facilitate consistent implementation of research security requirements across institutions

Goal:
Facilitate efficient navigation and implementation of evolving research security policies and subsequent compliance activities for all institutions, regardless of existing resources.

Approach:
Leverage the NSF-funded SECURE Center and its shared virtual environment to provide resources codesigned and developed with the regulated community, to facilitate consistency across institutions in the implementation of federal requirements, and to serve as a hub for research security information, communications, and resources. This option proposes using the SECURE Center as a central information hub for research security information and resources for assessing and mitigating risks to research, addressing federal requirements, and adapting the shared virtual environment to provide a clearinghouse for federal and institutional research security-related policies and processes, checklists, FAQs, and training modules among other resources. Using innovations such as artificial intelligence, this clearinghouse would incorporate interactive dashboards and include querying tools to facilitate better navigation of the collected information and identify updates to policies and procedures. SECURE Analytics, also part of the SECURE Program, can be leveraged to assess risks and mitigate and prevent foreign interference in research.

Pros:	Cons:
• Provides a central conduit and database of information and resources that all institutions can access. • Helps facilitate consistency in understanding and approaches to compliance. • Facilitates engagement and information sharing among institutions. • Uses a community-based approach to codesign and develop resources that meet the community's stated needs. • Provides risk analysis tools, resources, and reports to manage, mitigate, and prevent foreign interference.	• Without active agency engagement, the platform would be overly reliant on crowdsourcing information. Agency engagement could occur via the U.S. Government Steering Committee. • The SECURE Program is still a newer model and more information on its impact is needed. • To be successful, an adequate source of funding would be required to maintain and continue to modernize the platforms and tools. • A pay-for-use model will prohibit the engagement of under-resourced institutions. Federal funding is needed to avoid unintended inequities.

TABLE 2-7 Continued

Option 4.3: Amend the CHIPS and Science Act[b] to allow for "just-in-time" research security training

Goal:
Reduce the administrative workload for researchers on unfunded proposals, increase efficiency during the early grant process, and harmonize training requirements across funding agencies.

Approach:
DOE, the National Institutes of Health, and other Public Health Service federal funding agencies require that all principal investigators and other senior and key personnel complete COI training prior to expending project funds, and training is required every 4 years. Consistent with this, Congress could introduce amendments to the CHIPS and Science Act to allow for research security training at the time of award and to provide flexibility in the frequency of training. Clear guidance on training requirements could be communicated across federal agencies and the academic community to ensure the broad agency implementation of the just-in-time mechanism. To reduce the annual training hours required by current policies, agencies could subscribe to the SECURE Center's condensed research security training module and broadly accept and apply an investigator's completion of said module as satisfying the agency's security training requirement.

Pros:	Cons:
• Focuses training time to the period when researchers would most benefit from the training. • Mirrors existing mechanisms used by funding agencies, increasing consistency in pre-award compliance operations. • Allows institutions to allocate pre-award resources more efficiently.	• Requires congressional action.

Option 4.4: Establish common principles for agency risk reviews

Goal:
Retain agency variance resulting from differences in mission—for example, basic fundamental research versus research involving critical technologies and military applications—while effectively reducing inconsistencies in the application of risk reviews and mitigation across the government, reducing uncertainty, and providing clarity to risk review determinations and broader international engagement.

continued

TABLE 2-7 Continued

Approach:
Given the differing missions and research portfolios among federal research funding agencies, the procedures for risk reviews for fundamental research proposals result in inconsistent risk interpretation and mitigation. This leads to confusion among researchers and institutions with respect to international research engagement. To address these inconsistencies and the opacity of government actions, while also maintaining flexibility to account for their different missions and research portfolios, all agencies could adopt a standard set of fundamental principles and resources when conducting risk reviews and mitigation actions on federally funded research, both basic and applied. For example, agencies would use the same U.S. restricted party lists, assess use of education origins as a risk factor, assess a prohibition on use of non-U.S. risk tools, and use common principles and processes for risk mitigation. In addition, all agencies should develop a shared set of expectations for risk-mitigation issues identified in reviews with centralized tools to assist in harmonizing actions on areas of risk reviews.

Pros:	Cons:
• Increases the predictability of risk reviews and mitigation actions conducted by government agencies and facilitates a better understanding of acceptable engagement of international students and scholars. • Allows for greater consistency in how policies and procedures are applied.	• For proper uptake, federal research funding agencies will need to come to a consensus on common principles and mitigation procedures, which may take time, resulting in prolonged uncertainty on the rules of international engagement. • Will require oversight to ensure proper harmonization.

[a] NSF (U.S. National Science Foundation). n.d.-b. *NSPM-33 implementation guidance.* https://www.nsf.gov/bfa/dias/policy/nspm-33-implementation-guidance (accessed July 15, 2025).

[b] *Creating Helpful Incentives to Produce Semiconductors for America (CHIPS) and Science Act of 2022*, Public Law 117-167 (August 9, 2022).

Export Controls

The federal government has an extensive export control regulatory regime to protect U.S. trade and national security (NASEM, 2009, 2022). Multiple federal departments oversee export control regulations, and this is an area that has long needed regulatory reform. Academic institutions have experienced significant challenges fully adopting existing federal export control framework requirements given the expansive research areas within academia that require a broad knowledge of regulations, compared to industry where organizations focus on a smaller number of technologies. To meet the demands of complying with both export controls and research security regulations and requirements, institutions have had to find ways to identify and coordinate resources needed to increase their efforts for coming into compliance and centralizing activities within the institution (COGR, 2022). The specialized staff, software, and legal services needed to ensure compliance may be difficult for less-resourced campuses to afford. Universities also navigate potential conflicts between open science practices for fundamental research and export controls. While the results of fundamental research are excluded from export controls, tangible items and software which are not intended for publication are subject to controls.

U.S. export control compliance for universities is characterized by both structural fragmentation and doctrinal ambiguity. Three primary regulators maintain separate country- and item-specific controls: the U.S. Department of State, which administers International Traffic in Arms Regulations and the United States Munitions List (USML); the U.S. Department of Commerce, which administers the Export Administration Regulations and the Commerce Control List (CCL); and the U.S. Department of the Treasury, which administers Office of Foreign Assets Control and the Specially Designated Nationals and Blocked Persons List. Other agencies can also impose additional or emergency measures, such as the Federal Emergency Management Agency's pandemic-era restrictions on medical personal protective equipment (FEMA, 2021). Each agency follows distinct procedures for registrations, license determinations, exemptions, and end-use verifications, so an academic project containing multiple types of research may be subject to multiple overlapping requirements with no single point of harmonization.

Although the Fundamental Research Exclusion[18,19] is meant to remove export controls when the results of research are intended for unrestricted publication and open use, physical tools, equipment, software, and proprietary data used in such research must be treated as controlled items as applicable. Universities, therefore, must operate comprehensive export control programs that include screening, licensing, training, and controls—including expensive modifications to facilities—even when the underlying scholarship is exempt. The interplay of multiple agency rules and the partial reach of the exclusion creates a complex, resource-intensive compliance environment that can delay collaborations, increase administrative costs, and discourage international partnerships essential to scientific progress.

Striving to stay compliant, institutions sometimes employ their own additional policies or processes to preemptively address federal concerns—such as export control policies related to foreign nationals—that may further increase administrative workloads without demonstrably improving protection. As an example, while most H-1B visa holders pursuing training as postdocs or positions as research personnel at U.S. institutions conduct fundamental research, which is exempt from export controls, universities have implemented additional reviews and are required to submit I-129 petitions to confirm that foreign nationals are, in fact, exempt (Decrappeo et al., 2011). This additional review requires not only coordination and processing through human resources and international affairs offices, which are typically engaged in H-1B applications, but also sponsored research offices, compliance officers, technology licensing, and sponsoring faculty to appropriately complete I-129 petitions. This level of extensive review and processing is both onerous and provides only a snapshot of the research connected to the individual, which may change and then require additional processing. This overcomplicates petitioning for administrative personnel, sponsoring faculty, and visa petitioners.

The federal government has attempted some broader efforts at reform of export controls, such as the Export Control Reform Initiative that ran from 2009 to 2016 (DOS, 2013). Of particular focus for this initiative was an effort to move appropriate items from the USML with its more stringent regulations to the CCL, which allows easier export of items that

---

[18] The U.S. White House. *National Policy on the Transfer of Scientific, Technical and Engineering Information*, National Security Decision Directive (NSDD) 189. Washington, DC: Office of the President, September 21, 1985. https://irp.fas.org/offdocs/nsdd/nsdd-189.htm.

[19] "*Technology" or "Software" That Arises During, or Results from, Fundamental Research*, 15 C.F.R. Part 734.8 § 734.8 (March 25, 1996).

OPTIONS TO OPTIMIZE THE RESEARCH ENTERPRISE

do not require greater control (Stricker and Albright, 2017). Over the course of the initiative, all categories in the USML were reviewed, except for firearms, ammunition, and artillery, to consider whether they could be moved to the CCL (Insinna, 2017). Other outcomes of this effort include a license exception called a Strategic Trade Authorization to facilitate export transfer to low-risk countries and the creation of an Export Enforcement Coordination Center to coordinate export control enforcement across agencies (CRS, 2020).

TABLE 2-8 Options to Address Export Controls (Regulatory Area 4)

Option 4.5: Renew the Export Control Reform Initiative[a] with input from academia

Goal:
Continue prior efforts to streamline and clarify export controls and reduce associated administrative work with representation from the academic research community.

Approach:
The Export Control Reform Initiative began in 2009 with the aim of streamlining export controls by simplifying processes and increasing coordination across agencies, enhancing the clarity of descriptions of controlled items, and transferring appropriate items from the United States Munitions List to the Commerce Control List to reduce the level of control for those items. Congress and the White House could initiate another reform campaign to continue and expand upon some previous successful efforts and include representation from the academic research community.[b]

Pros:	Cons:
• Builds and expands upon prior success.	• Engaging agencies in coordinated reform efforts can be challenging. Staffing shortages could contribute to this challenge.
• Enables consideration of research when streamlining processes and establishing controls.	

Option 4.6: Adopt a risk-tiered approach to export controls

Goal:
Reduce unnecessary work on low-risk research, transporting information, technology, or other research-related items of U.S. national interest overseas.

Approach:
Export controls are broadly applied for industrial purposes, resulting in controls that are inconsistent with the academic research use of items, services, and technologies. Instead, barriers could be removed for those engaged in the lowest-risk work and appropriately tiered for others. The Fundamental Research Exclusion[c] does not include the technology or tools used to conduct research. The Departments of State and Commerce could allow for greater risk-tiered variation in requirements and regulations for their controlled items, particularly for research covered under the Fundamental Research Exclusion.

continued

TABLE 2-8 Continued

Pros:	Cons:
• More time and attention could be given by both researchers and those overseeing regulations to the highest-risk work. • Freer flow of knowledge informing and being produced by fundamental research to benefit from desirable, low-risk international research collaborations.	• Additional efforts by the two agencies are needed to determine risk-tiered controls for all or most controlled items.

Option 4.7: Expedited licensing processes for low-risk controlled research and transparency and clarity in Office of Foreign Assets Control (OFAC) license processes

Goal:
Expedite export-control licensing requests using a risk-tiered, fast-track licensing pathway for low-risk controlled research.

Approach:
Lengthy licensing processes are a particular hindrance for low-risk but controlled research. Currently, the licensing process is not risk-tiered, but rather a first-come, first-served process, which can create delays to projects that present minimal security concerns, taking weeks or months to process and slowing the pace of research. Processes could be developed for low-risk research, similar to the ones undertaken in the voluntary self-disclosure program within the Department of Commerce, that reduce and fast-track licensing processes. In addition, OFAC licensing can take as long as a year. Greater transparency and clarity regarding the process would be beneficial.

Pros:	Cons:
• Reduces administrative delays, frees up agency resources, and allows researchers to begin work sooner.	• Requires additional effort by agencies to develop an appropriate expedited licensing process for low-risk research controls.

[a] The White House. 2013. *Fact sheet: implementation of export control reform.* https://obamawhitehouse.archives.gov/the-press-office/2013/03/08/fact-sheet-implementation-export-control-reform (accessed August 12, 2025).

[b] CRS (Congressional Research Service). 2019. *The U.S. export control system and the Export Control Reform Initiative.* https://www.congress.gov/crs_external_products/R/PDF/R41916/R41916.46.pdf (accessed July 1, 2025).

[c] BIS (Department of Commerce Bureau of Industry and Security). 2011. *Deemed exports and fundamental research for biological items.* https://www.bis.doc.gov/index.php/2011-09-08-19-43-48 (accessed August 12, 2025).

Cybersecurity and Data Management

The federal data governance and cybersecurity landscape for academic research is a patchwork of overlapping and sometimes contradictory requirements. The National Institute of Standards and Technology is the technical backbone of federal cybersecurity policy and practice and has overall responsibility for developing cybersecurity standards, guidelines, best practices, and frameworks. Federal agencies use these policies, yet many, such as NIH, NSF, DOE, DOD, and USDA, impose agency-specific policies on data sharing, stewardship, privacy, and security. Application of the terms *Controlled Unclassified Information (CUI)*, *federal contract information*, and *fundamental research* is not consistent across agencies, or, as in a recent NIH policy change, CUI data protection standards are implemented for data that the agency acknowledges are not CUI. CUI is often applied in agency agreements and contracts where the proposed work is fundamental research. Monitoring and enforcement are uneven, and many rules have not kept pace with evolving scientific practice.

Agencies are applying different standards for handling and storing CUI, and some agencies struggle internally to apply applicable controls in a consistent manner. Research institutions must maintain parallel compliance programs, duplicate training, and multiple documentation streams to satisfy incompatible requirements, yet they still lack an authoritative source that reconciles definitions, risk categories, templates, and reporting formats. In the absence of a coordinated cross-agency framework, researchers and institutions face shifting guidance, redundant oversight, and significant administrative costs, while the nation falls short of delivering consistent and risk-appropriate protection for federally funded data. The sheer number of categories makes CUI an especially challenging issue.

TABLE 2-9 Options to Address Cybersecurity and Data Management (Regulatory Area 4)

Option 4.8: Provide clarity for definitions on Controlled Unclassified Information (CUI) and fundamental research

Goal:
Increase clarity on policies related to CUI.

Approach:
The White House could expand the definition of "fundamental research" beyond the scope of National Security Decision Directive 189 (NSDD-189)[a] to include all basic and applied research performed at U.S. institutions of higher education that is generally published and openly available for the scientific community.[b] In addition, the National Archives and Records Administration can provide clear guidelines for federal contract information[c] and CUI[d] with a definition of CUI to be used by all authoritative sources.

Pros:	Cons:
• More consistent and therefore effective application of CUI requirements.	• Changing agency practice and then ensuring that institutions are complying can be limited by the capacity of specific agencies. • Changing the long-standing definition of fundamental research could open up the broader question of whether NSDD-189 is still relevant today and whether fundamental research should be subject to national security controls.

Option 4.9: Develop a coordinated cross-agency framework for research, data security, and governance

Goal:
Align and streamline federal data governance requirements across agencies.

Approach:
The Office of Science and Technology Policy (OSTP) or National Institute of Standards and Technology (NIST) could coordinate a working group of the federal agencies with relevant policies to review and align existing data management, sharing, privacy, and security policies. The working group would develop a unified framework with standard definitions, risk categories, templates, and compliance expectations. Federal agencies could adopt the framework while allowing for mission-specific flexibility.

TABLE 2-9 Continued

Pros:	Cons:
• Reduces duplicative efforts across agencies and institutions. • Enhances interdisciplinary research, data sharing, and collaborative science. • Enables strategic national coordination.	• Risks of over standardizing and missing critical nuances for different science needs.

Option 4.10: Adapt cybersecurity requirements for university settings

Goal:
Align cybersecurity requirements for institutions of higher education to reflect the current open science environment that underpins U.S. innovation.

Approach:
The secretary of commerce could direct the under secretary of commerce for standards and technology and director of NIST, in collaboration with OSTP and the broader research community, to undertake a comprehensive review of cybersecurity controls as they apply to institutions of higher education and make appropriate modifications to ensure alignment with the academic research environment with attention to the growing adoption of an open science framework.

Pros:	Cons:
• Ensures that cybersecurity policies and practices, which were designed principally for government agencies and industry, are designed to meet the unique needs and largely fundamental research of higher education. • Reduces unnecessary time spent by higher education on applying controls that are inappropriate and unnecessary for its research and education mission.	• Institutions of higher education may need to discontinue some current practices and policies, which might cause confusion and create issues with other interrelated policies that need to remain in place.

[a] The White House. 1985. *National Policy on the Transfer of Scientific, Technical, and Engineering Information.* https://www.acq.osd.mil/dpap/dars/pgi/docs/National_Security_Decision_Directive_189.pdf (accessed July 11, 2025).

[b] COGR (Council on Governmental Relations). 2025. *Actionable ideas to improve government efficiency affecting the performance of research.* https://www.cogr.edu/sites/default/files/Actionable Ideas to Improve Gov Efficiency COGR_0.pdf (accessed July 8, 2025).

[c] *Basic Safeguarding of Covered Contractor Information Systems,* 48 C.F.R. Part 52, Subpart 52.2, § 52.204-21 (June 11, 2025).

[d] *Controlled Unclassified Information,* 32 C.F.R. Part 2002 (September 14, 2016).

REGULATORY AREA 5: RESEARCH INVOLVING BIOLOGICAL AGENTS

Multiple agencies oversee research involving biological agents, and there have been successful efforts to harmonize oversight over the years. Though these efforts are laudable, agencies continue to promulgate new requirements that continue to shift the burden of oversight to research institutions and individual principal investigators, often with insufficient guidance.

The U.S. Centers for Disease Control and Prevention (CDC)-NIH *Biosafety in Microbiological and Biomedical Laboratories* (BMBL) is generally accepted by federal agencies as the authoritative guidance document for biocontainment practices and laboratory safety (CDC, 2020). The *NIH Guidelines for Research Involving Recombinant or Synthetic Nucleic Acid Molecules (NIH Guidelines)* oversee recombinant pathogens and apply to research at NIH-funded entities. The *NIH Guidelines* govern research with synthetic and recombinant nucleic acid molecules and gene-drive modified organisms, establishing Institutional Biosafety Committees (IBCs) as the local-level oversight bodies that review and approve relevant research. While there are limits on the scope and applicability of the *NIH Guidelines* and BMBL, both are generally applied broadly and describe risk-tiered practices for safely conducting research with recombinant DNA and microbiological agents.

IBCs serve as the focal point for institutional biological safety oversight, and as new policies are issued, the work of IBCs has grown. The core function of IBCs is to oversee synthetic and recombinant nucleic acid molecule research, including gene-drive modified organisms, as described in the *NIH Guidelines*. However, IBCs at most institutions play a role in the oversight of a broader range of microbiological and recombinant DNA research than what is specifically articulated in the *NIH Guidelines* (Johnson and Dobos, 2019). They are also becoming increasingly involved in implementing policies around dual-use research of concern (DURC) and dangerous gain-of-function (GOF) research, representing an expansion of the scope of many IBCs from biosafety to biosecurity. DURC involves research that could be misused to threaten public health, agriculture, or national security. GOF studies, in this context, refer to research that enhances the pathogenicity or transmissibility of pathogens—particularly those with pandemic potential—to better understand their infectious nature. The May 5, 2025, EO *Improving the Safety and Security of Biological Research* directs agencies to revise or replace previous policies on DURC and GOF and aims to, among other things, stop dangerous GOF research, increase top-down oversight and accountability, and manage risks associated with non-federally funded research (CRS, 2025; The White House, 2025). At the time of this writing, DURC and GOF policy outcomes are uncertain, but in implementing this recent EO, federal agencies have begun asking research institutions to identify and report any "dangerous gain-of-function" research they are conducting, providing no additional guidance outside of the broad descriptive language in the EO.

The Federal Select Agent Program, administered jointly by CDC and USDA, oversees the possession, use, and transfer of select agents and toxins that pose a threat to public, animal, or plant health. Using a list-based approach, these regulations require added safety and security measures, registration, security screenings of individuals enrolled in the program, inspections, and more to help ensure that research involving high consequence pathogens is conducted to the highest standards of safety and security.

Biological agents and toxins are also covered under export control regulations with certain agents requiring controls on their possession, use, and transfer. Taken together, the oversight of biological agents and toxins involves multiple federal agencies and policies, each with varying jurisdictions, scopes, and applicability, and all seeking to manage the same or similar risks of inadvertent or intentional release of pathogens that could endanger the health of those working with these organisms and the public. While new policies have been implemented, there continues to be multiple oversight frameworks, insufficient guidance, and an increased reliance on institutions and individual principal investigators to identify research with potential national security implications or potentially "significant societal consequences" (The White House, 2025).

The committee provides options to harmonize these overlapping regulations through centralization, simplified guidelines, and reducing duplicative efforts.

Problem: Complex and overlapping regulations for research involving biological agents.

There are a number of overlapping regulations, guidelines, and policies from multiple federal agencies for the oversight of research involving biological agents and toxins. These can be complex, be redundant, and require institutions to implement multiple sets of rules that could have been integrated. In addition, researchers must also navigate export controls for the transfer and communication of information about some pathogens. These policies are promulgated by multiple agencies, and their scopes and applicability vary.

TABLE 2-10 Options to Address the Complex and Overlapping Regulations for Research Involving Biological Agents (Regulatory Area 5)

Option 5.1: Adopt a more centralized, more coordinated U.S. government-wide approach to regulating research involving biological agents and toxins

Goal:
Create a consistent and coordinated approach to reduce discrepancies and duplicative efforts across regulations of biological agents and toxins.

Approach:
The existing oversight frameworks provide a solid foundation for research involving biological agents and toxins but would benefit from a holistic effort to better centralize, coordinate, and clarify rules. A single agency, such as the National Institutes of Health (NIH) or another appropriate entity, could be charged with empowering and registering Institutional Biosafety Committees (IBCs) to provide oversight at the institutional level, regardless of the institution's federal funding source. That agency or entity could be the focal point for communicating with IBCs and could coordinate with federal partners as needed. The *Biosafety in Microbiological and Biomedical Laboratories* (BMBL) document[a] could remain as the authoritative guidance document describing containment and safety practices and gain wide adoption from federal agencies and research institutions. The Federal Select Agent Program[b] could continue to provide enhanced risk-tiered oversight of higher consequence pathogens and could work closely with the entity overseeing IBCs to harmonize guidance and requirements.

Pros:
- Builds on existing oversight frameworks, leveraging strengths.
- Centralizes and harmonizes primary regulations governing research with biological agents for all U.S. research, including private research.
- Simplifies implementation and oversight.

Cons:
- Microbiological research is inherently dynamic and risks change as new microbes are discovered and countermeasures developed. Therefore, guidance needs to be regularly updated to ensure oversight remains appropriately risk based. This creates additional challenges for coordination, as the IBC would need to continually coordinate with agencies to update guidance as the research evolves.
- Would still require multiple oversight frameworks, and since federal agencies have authority over and responsibility for the research they fund, they may struggle to centralize IBC registration or harmonize oversight sufficiently.

TABLE 2-10 Continued

- Requires major legislative and executive action, particularly if oversight is intended to cover non-federally funded research.

Option 5.2: Simplify and harmonize current NIH/U.S. Department of Agriculture/U.S. Centers for Disease Control and Prevention guidelines, and exempt low-risk activities

Goal:
Ensure a risk-tiered, harmonized approach to guidelines for biological agents and toxins.

Approach:
The *NIH Guidelines* and BMBL provide risk-tiered guidance and oversight, but there are still examples of low-risk research that continue to require more stringent oversight, requiring researchers and institutions to devote unnecessary administrative time to activities that do not appreciably improve safety. NIH could review the current *NIH Guidelines* in an effort to improve clarity and remove unnecessary oversight requirements for low-risk recombinant and nonrecombinant activities. The Federal Select Agent Program could do likewise, continuing its periodic evaluations of the lists of covered agents. This approach would reduce administrative workload by removing review, approval, and reporting requirements for lower-risk research but would not be intended to increase safety risks as researchers would still be expected to follow general safety procedures such as wearing suitable personal protective equipment.

Pros:	Cons:
• Builds on existing regulatory documents and leverages existing mechanisms to simplify and carve out exemptions for lower-risk activities.	• Still would require multiple regulations and guidance.

Option 5.3: Largely remove biosecurity and national security oversight from the purview of IBCs and focus the oversight of dual-use research of concern (DURC) and dangerous gain-of-function (GOF) research at the federal level

Goal:
Streamline the oversight of DURC and GOF to ensure the most equipped bodies are focused on this.

continued

TABLE 2-10 Continued

Approach:
DURC and GOF policy efforts have focused on policy concerns around the potential misuse or inadvertent release of the highest-consequence pathogens. IBCs could continue to play a role in the local-level biosafety oversight of such work, but they are not well equipped to make determinations involving national security. Such determinations require expertise in national intelligence, security and vulnerability assessments, public health preparedness, and other areas that are within the purview of the federal government but not universities or research institutions. Rather than relying on IBCs and institutions to screen their portfolios or build new oversight mechanisms around research activities that are difficult to define or assess, federal funding agencies could identify, at the time of funding, whether any studies meet their definitions of DURC or dangerous GOF research. Then, the funding agency could enter into a dialogue with the institutions, prior to beginning the research, to collaboratively identify appropriate conditions and containment measures for safely conducting and communicating the research. Institutions could implement agreed-upon requirements for safely conducting the research. The Office of Science and Technology Policy (OSTP) could also serve as a coordinating function and a venue for sharing determinations by the agencies to ensure consistency. In cases where an agency may be unsure of whether to fund a certain study or is unclear on how best to manage risks, OSTP could convene experts from across the government to advise the agency. The federal government could continue to provide guidance to institutions about DURC and GOF, and could encourage communication in the event of questions, concerns, or unexpected research developments, but the primary oversight role would reside with the federal funders.

Pros:	**Cons:**
• Recognizes the differing expertise and strengths of federal agencies and research institutions and calls on each to operate within their appropriate capacities. • Removes an oversight responsibility from researchers and research institutions that was confusing and onerous and had little added benefit.	• Research is dynamic and unanticipated results could, in fact, generate biological agents, information, or technologies that could be misused to cause harm. While unlikely, under this option institutions would have less responsibility for identifying these cases.

[a] HHS (U.S. Department of Health and Human Services), CDC (U.S. Centers for Disease Control and Prevention), and NIH (National Institutes of Health). 2020. *Biosafety in microbiological and biomedical laboratories (BMBL), 6th Edition.* https://www.cdc.gov/labs/pdf/SF__19_308133-A_BMBL6_00-BOOK-WEB-final-3.pdf. (accessed August 12, 2025).

[b] HHS and USDA (U.S. Department of Agriculture). n.d. *Federal select agent program.* https://www.selectagents.gov/ (accessed August 8, 2025).

[c] NIH (National Institutes of Health). 2024c. *NIH guidelines for research involving recombinant or synthetic nucleic acid molecules.* Office of Science Policy, U.S. Department of Health and Human Services. (Amendments effective September 30, 2024).

REGULATORY AREA 6: HUMAN SUBJECTS RESEARCH

There has been some progress in reforming the regulatory framework for overseeing human subjects research in the United States, but the system remains fragmented. While reforms over the past decade have aimed to modernize oversight, reduce administrative burden, and better align protections with current research practices, implementation has been inconsistent and has sometimes introduced new challenges. Many federal requirements remain siloed or are not well coordinated across agencies. As with other areas of regulations, these disconnects create inefficiencies in how institutions and agencies apply oversight requirements, which can undermine the ethical and scientific integrity the system is meant to support.

The principles and guidelines governing the protection of human research subjects in the United States date to the 1979 Belmont Report, issued under the National Research Act of 1974[20] to ensure all such research adhered to three basic ethnical principles: respect for persons, beneficence, and justice (HHS, 2018, 2024b). Today, human subjects research is more directly guided and regulated by the Federal Policy for the Protection of Human Subjects, also known as the Common Rule, because it is codified in separate regulations by each of its agency and department signatories.[21] It is ultimately at the discretion of each federal agency as to whether or not the Common Rule applies. While 20 agencies are current signatories to the Common Rule (HHS, 2025), not every agency that funds human subjects research has signed on.

Adoption of the Common Rule across agencies created needed harmonization, but there remain challenges in creating and implementing a fully cohesive policy for protecting human subjects across the federal government. This is seen, for example, in inconsistent definitions across agencies for key terms such as "clinical trial," "human subjects," and "engagement in research," as well as duplicative or inconsistently applied reporting requirements, such as variations in ClinicalTrials.gov registration or adverse event reporting.[22]

Another challenge is inconsistencies in implementation and interpre-

---

[20] *National Research Act of 1974,* Public Law 93-348, 93rd Congress, 2nd sess. (July 12, 1974).

[21] *Federal Policy for the Protection of Human Subjects,* 82 Fed. Reg. 7149-7274 (January 19, 2017).

[22] For more information about relevant human subjects research definitions (and differences between FDA and NIH), please see Comparison of FDA and HHS Human Subject Protection Regulations | FDA.

tation that contribute to regulatory fragmentation. For example, research funded by HHS is subject to additional regulatory protections for certain populations of research subjects beyond the Common Rule,[23] while DOD adds additional unique restrictions on waivers of informed consent,[24] and DOE has specific requirements for protecting personally identifiable information.[25] These differences can increase administrative complexities, delay study start-up, and lead to variability in how regulatory requirements are applied by researchers and IRBs.

Clinical investigations provide another set of examples of human subjects research facing particular challenges under the Common Rule. Human subjects research encompasses a variety of different research, including social and behavioral research and medical or clinical research, with varying degrees of regulation based on risk. Clinical investigations understandably require greater oversight to ensure participant safety but are subject both to policies from agencies that have adopted the Common Rule, such as HHS, DOD, VA, and others, and FDA regulations that are similar but not identical to the Common Rule, resulting in unnecessary duplicative oversight. Clinical trials also face potential conflicting requirements between the Common Rule and the Health Insurance Portability and Accountability Act (HIPAA) regulations for protected health information (PHI). HIPAA protects the privacy and security of identifiable health information held by covered entities, requiring specific authorizations or waivers for its use, while the Common Rule governs the ethical conduct of human subjects research through informed consent, IRB oversight, and safeguards for vulnerable populations. Their requirements often diverge—including differing definitions of "identifiable" data and "de-identification" and slightly different criteria for waivers of consent or authorization. This means a dataset could be considered nonidentifiable under the Common Rule but still be PHI under HIPAA, which means that researchers are then required to meet two separate standards for the same project.

Another significant challenge for human subjects research exists for work conducted across multiple sites. The 2016 National Academies' report recognized that additional work and burden is created in multisite studies when each site needs to get approval of their materials and procedures from their individual IRBs. To streamline the processes, that report recom-

---

[23] *Protection of Human Subjects,* 45 C.F.R. Part 46 (August 21, 2025).

[24] *Protection of Human Subjects and Adherence to Ethical Standards in DOD-Conducted and –Supported Research,* DOD Instruction 3216.02 (June 29, 2022).

[25] *Protection of Human Research Subjects,* DOE 0 443.1C Chg.1 (November 26, 2019).

mended creating a single IRB on which all sites would rely. NIH adopted this recommendation in 2016, with required implementation by 2018. Use of a single IRB for multisite studies was additionally adopted in the 2018 revisions to the Common Rule, with slated implementation by 2020.

The 2018 Common Rule revision introduced new exemption categories regarding benign behavioral interventions and secondary research involving identifiable private information or biospecimens when broad consent is obtained, as well as modernized consent forms (HHS, 2017). At the same time, the revised Common Rule now includes the concept of "limited IRB review" for some categories that were previously exempt, such as research involving surveys of benign behavioral interventions and storage or maintenance of identifiable private information or biospecimens for secondary research use.

While this broadened the type of research that qualifies for exemption, it also introduced a new category of review requiring the development of new forms and procedures for IRBs that increased burden. Since adoption, challenges and unintended consequences have arisen that have not reduced duplicative reviews and that need to be addressed to ensure the single IRB does in fact reduce burden (Cathrall, 2018). For example, while the single IRB was intended to prevent the need for IRB approval at each individual site, in practice, some institutions have been reluctant to rely on a single IRB of record or have faced challenges navigating this change, such as resistance to change as well as continuing institutional oversight responsibilities (Burr et al., 2022; Green, et al., 2023).

Similarly, Common Rule revisions that expanded the types of exempt studies should mean that such studies are not included in protocol data provided to the Office for Human Research Protections. However, IRBs have not implemented these revisions consistently, resulting in potential discrepancies and inaccuracies across institutions (GAO, 2023). With the lack of harmonization and hesitancy to centralize processes, IRB workloads in some cases have increased significantly, and Common Rule implementation has been slowed across agencies (Green et al., 2023).

Beyond the Common Rule itself, institutions must also navigate an expanding array of intersecting federal requirements not originally designed for human subjects research oversight but that intersect with this oversight, such as data security, research security, export controls, and privacy regulations. For example, export control and research security regulations might apply for a study involving human subjects that includes foreign collaborations, data sharing, or international travel. IRBs typically do not have expertise in these areas, but decisions about study approval may hinge on whether potential risks have been properly assessed. The absence of an appropriate framework for integrating requirements like this into an existing research oversight structure such as a Human Research Protection Program adds further complexity to the research process and can lead to inconsistent or duplicative compliance approaches.

Finally, challenges in human subjects research occur as a result of a lack of practical flexibility in the regulatory framework and inadequate adaptation to evolving research methods and technology (Ehidiamen and Oladapo, 2024; Fleischman, 2005; Walch-Patterson, 2020). With limited flexibility in practice, requirements and regulations can face unnecessary delays and are still insufficiently calibrated to risk. Without appropriate adaptation to new methods and clear guidance on integrating outside requirements, there is uncertainty in how to ensure compliance when engaging new and innovative research practices as well as when dealing with intersecting regulations.

The committee proposes options to streamline regulations, consider revisions to the single IRB structure, and develop risk-based tiered and modernized approaches to human subjects research.

Problem 1: Continued variation across agencies in human subjects regulations, oversight, reporting, and definitions.

Even with the adoption of the Common Rule as the overarching framework for human subjects research, the complexity and overlap of multiple regulations and agency requirements can create significant administrative burdens for researchers and staff. Navigating state, federal, and agency requirements, as well as understanding which regulations apply to specific research, can lead to confusion and delays in the approval process. Specific challenges arise for clinical investigations, for which both regulations from agencies that have adopted the Common Rule and FDA regulations can apply, resulting in duplicative efforts. Problems also arise when HIPAA regulations and the Common Rule both apply but may conflict, such as in instances when differences in definitions lead data to be considered nonidentifiable under the Common Rule but PHI under HIPAA. In addition, variation in the definitions of key terms and reporting requirements creates unnecessary uncertainty and lack of clarity for researchers.

TABLE 2-11 Options to Address the Continued Agency Variation in Human Subjects Regulations (Regulatory Area 6)

Option 6.1: Establish a federal harmonization mechanism with joint agency guidance

Goal:
Provide consistent guidance, reduce regulatory duplication, and maintain alignment in human subjects research oversight.

Approach:
To address fragmentation and confusion in human subjects research oversight, the federal government could establish a mechanism, such as an interagency working group under the Office of Science and Technology Policy or the Office of Management and Budget's Office of Information and Regulatory Affairs, and possibly using the National Science and Technology Council, with clear authority and accountability to align human subjects research policies, definitions, and review processes across agencies and maintain ongoing coordination. This body could be empowered to review proposed and existing regulations for consistency, recommend policy adjustments to reduce duplication, and monitor agency implementation. This body could also lead a coordinated effort to review and align key definitions relevant to human subjects research across federal agencies, including "clinical trials," "human subjects," and "engagement in research" with enforcement capability to ensure agency participation. As a first step, participating agencies could issue joint federal guidance clarifying the applicability of overlapping U.S. Food and Drug Administration (FDA) and Common Rule[a] requirements, including practical examples outlining when each framework applies and clear compliance pathways for dual-regulated studies. As an ongoing effort, this body could regularly solicit feedback from the regulated community to identify additional problem areas related to any joint agency guidance and effect solutions to address any issues. This approach provides immediate clarity to researchers while ensuring long-term consistency and enforceable alignment across agencies.

Pros:
- Provides actionable clarity for institutions navigating overlapping regulations.
- Establishes accountability for ongoing federal coordination.
- Prevents future fragmentation as new policies are developed.
- Supports consistent federal communication with stakeholders.
- Enhances efficiency in compliance and review processes.

Cons:
- Requires agency participation, commitment, and willingness to adjust practices.
- May require statutory or regulatory support to ensure authority for implementation or modification of existing regulations, a time-consuming process.
- Needs dedicated administrative resources to maintain alignment and track progress.

continued

TABLE 2-11 Continued

Option 6.2: Lead a cross-agency review to streamline and align federal human subjects research requirements

Goal:
Reduce complexity and duplication in federal human subjects research regulations while maintaining necessary protections.

Approach:
The U.S. Department of Health and Human Services (HHS) and other federal agencies that conduct or support human subjects research could lead a comprehensive cross-agency review of federal human subjects research regulations to identify duplicative, outdated, or conflicting requirements. This review could engage all agencies that have signed on to the Common Rule, FDA, and relevant HHS offices and incorporate stakeholder input with the goal of developing a roadmap for regulatory alignment and simplification while maintaining participant protections. This effort could also examine agency-specific processes, including separate human subjects review and/or extra layers of oversight requirements imposed by some agencies, which have been cited by researchers and staff who support their work as duplicative.

Pros:
- Allows a reexamination of existing requirements to determine whether they continue to provide meaningful participant protections, recognizing that needs and contexts may have changed over time.
- Reduces unnecessary regulatory complexity.
- Clarifies agency requirements for institutions and researchers.
- Allows institutions to redirect resources toward research.
- Reinforces public trust by demonstrating federal commitment to reviewing and updating oversight systems to ensure they remain effective, relevant, and focused on meaningful protections.

Cons:
- Requires sustained interagency commitment and resources.
- May face resistance from agencies without a clear mandate.
- Could require statutory modification and/or changes to regulations for complete alignment.

Option 6.3: Issue joint federal guidance clarifying definitions and application

Goal:
Provide immediate, actionable clarity while longer-term regulatory alignment efforts proceed.

TABLE 2-11 Continued

Approach:
Federal agencies could collaboratively issue joint guidance clarifying the definitions and applications of key terms within human subjects research regulations and agency guidance. As an example, definitions of "identifiable" and "de-identification" differ between the Common Rule and Health Insurance Portability and Accountability Act (HIPAA) regulations, and this and other conflicts could be addressed through ensuring consistency in terms. This guidance could use practical examples and case studies to illustrate how definitions apply across various research contexts, reducing confusion and variability in interpretation while agencies work toward longer-term harmonization.

Pros:
- Can be implemented relatively quickly.
- Provides practical tools for institutions and researchers.
- Reduces near-term confusion while maintaining flexibility for agencies.

Cons:
- May not fully resolve underlying definitional inconsistencies.
- Requires sustained agency cooperation to remain current and effective.

Option 6.4: Develop a centralized federal reporting framework

Goal:
Reduce duplicative and inconsistent federal reporting burdens while maintaining transparency and accountability in human subjects research.

Approach:
Federal agencies, led by HHS along with other agencies that conduct or support human subjects research, could collaborate to establish a unified reporting framework that standardizes human subjects research reporting requirements across agencies while eliminating duplicative submissions. This could involve creating a shared reporting portal for investigator activities such as trial registration, results reporting, and adverse event submissions, enabling institutions to meet obligations across agencies through a single consistent process. This could also work to standardize systems across agencies for Institutional Review Board (IRB) reporting requirements such as determinations of suspension, termination, and unanticipated problems.

Pros:
- Simplifies reporting processes for institutions and investigators.
- Reduces administrative workload and compliance confusion.
- Maintains transparency and participant protection oversight.

Cons:
- May require updates to statutes, regulations, or agency policies for full implementation.
- HHS may not subscribe to leading the effort or have the ability to enlist others.
- Agencies may resist changes that reduce individual control over reporting systems.

continued

TABLE 2-11 Continued

	• Institutions may need to adjust internal systems and workflows during the transition. • Requires sustained federal investment and maintenance of shared infrastructure.

Option 6.5: Conduct a federal review of reporting requirements and issue joint guidance

Goal:
Eliminate redundant reporting, reduce unnecessary administrative workload, and provide clear, consistent guidance while maintaining transparency and participant protections.

Approach:
HHS could lead a comprehensive review of human subjects research reporting requirements across federal agencies, including FDA, the U.S. Department of Defense, U.S. Department of Veterans Affairs, U.S. National Science Foundation, and others involved in overseeing human subjects research, to identify redundancies and opportunities for streamlining. Following this review, agencies could issue joint guidance clarifying which reporting requirements for investigators as well as for IRBs are necessary, aligning definitions and timelines, and providing clear examples to reduce duplicative reporting and compliance confusion.

Pros:	Cons:
• Promotes evidence-based refinement of reporting policies. • Provides immediate clarity for institutions and investigators. • Encourages alignment of agency expectations and timelines. • Demonstrates federal commitment to reducing unnecessary burden.	• Agencies may be resistant to harmonizing requirements due to differing missions or priorities. • HHS may not subscribe to leading the effort or have the ability to enlist others. • May require statutory, regulatory, or policy changes for full alignment. • Implementation could take time, delaying immediate burden reduction. • Institutions may need to adapt procedures as guidance and definitions are updated.

Option 6.6: Make FDA the sole regulatory agency for human subjects research for clinical investigations[b]

TABLE 2-11 Continued

Goal:
Streamline regulations for clinical investigations and reduce redundant and duplicative regulatory oversight.

Approach:
The FDA, which oversees clinical investigations, is not a Common Rule agency. While required to harmonize with the Common Rule to the extent possible by law, it ultimately has differing regulations.[a] Rather than continue duplicative efforts across both the Common Rule and FDA, FDA would be established as the sole regulatory agency for clinical investigations. Researchers conducting clinical investigations would be subject only to FDA rules and requirements.[c]

Pros:	Cons:
• Reduces duplicative oversight under FDA and the Common Rule. • Ensures one consistent set of rules is governing all clinical investigations. • Reduces confusion for researchers navigating human subjects regulations for clinical investigations.	• Does not bring FDA requirements, which are more strict in some instances, such as adding additional IRB requirements for investigational new drugs, but less strict in others, such as providing additional protections for vulnerable populations, into full alignment with other agencies applying the Common Rule. • Does not address duplicative requirements for other agencies. • May create challenges in areas, such as the use of artificial intelligence (AI), which are not consistently interpreted across IRBs as medical devices constituting a clinical investigation. IRBs may differ in these instances whether they defer to FDA or HHS regulations.

Option 6.7: Centralize compliance information in one accessible platform

Goal:
Simplify access to regulatory information, making it easier for the various stakeholders to understand which regulations apply to their specific research projects and reduce time spent navigating multiple sources, and improve consistency in compliance across the institution.

continued

TABLE 2-11 Continued

Approach:
This solution involves creating a centralized digital platform where all relevant regulatory and agency requirements are stored in one place so that institutions and other organizations can access the most up-to-date requirements from a centralized location. Researchers and institutional staff could access these regulations through an intuitive search function or browse through categories specific to their type of research. HHS could lead this effort in partnership with other federal agencies and allow states and institutions to build out information on further regulations/requirements for researchers to be able to navigate. This effort could leverage work in the International Compilation of Human Research Standards[d] as well as similar compilations in other areas, such as privacy and research security.

The platform would feature a user-friendly interface where users can search for specific regulations by keywords, research type, or regulatory category. It could also include helpful links to guidance documents and FAQs. The system would be continuously updated as new regulations are released, ensuring that researchers always have access to the latest compliance information.

Pros:	Cons:
• Reduces time spent identifying and navigating complex regulatory requirements.	• Development and ongoing maintenance can be resource intensive and costly.
• Provides quick and clear access to necessary guidelines.	• The complexity of the various regulatory requirements and understanding how these intersect requires specific levels of skill and continuous oversight.
• Improves overall efficiency by having all compliance information in one location.	• HHS may not subscribe to leading the effort or have the ability to enlist others.

Option 6.8: Implement an interactive decision-support tool or flowchart for determining applicable regulations

Goal:
Enable researchers and staff to more quickly and accurately identify compliance requirements earlier in the process.

TABLE 2-11 Continued

Approach:
This solution involves creating an interactive decision-support tool or flowchart that guides researchers through a series of questions to determine which regulations apply to their specific research project. Researchers could input key details about their study, such as the study type, research methodology, participant population, and funding sources, and the tool would generate a list of relevant regulatory and compliance requirements. Spearheaded by HHS and in partnership with other federal agencies, the tool could function as an interactive, step-by-step decision tree. Researchers would answer a series of yes or no questions or select options that describe their study. Based on these inputs, the tool would suggest a list of applicable regulations, such as the Common Rule, HIPAA,[e] FDA regulations, or other relevant laws, regulations, and agency guidance. It could also provide links to regulatory documents or guidance materials that researchers can review for further details. This tool could draw on AI tools to aid in producing suggestions for applicable requirements.

Pros:	Cons:
• Provides quick and clear answers on which regulations apply to a given study.	• Limited complexity for more nuanced or interdisciplinary research, which might require manual interpretation.
• Easy to use for researchers, reducing the need for manual searching through complex regulations.	• Regular updates will be required to ensure the tool stays current with evolving regulations.
• Customizable to various study types—allowing for tailored compliance recommendations.	• Initial development can be resource intensive to build the decision logic and ensure the tool's accuracy.

[a] HHS (U.S. Department of Health and Human Services). 2025. *Federal policy for the protection of human subjects ("common rule")*. https://www.hhs.gov/ohrp/regulations-and-policy/regulations/common-rule/index.html (accessed July 8, 2025).

[b] COGR (Council on Governmental Relations). 2025. *Actionable ideas to improve government efficiency affecting the performance of research*. https://www.cogr.edu/sites/default/files/Actionable Ideas to Improve Gov Efficiency COGR_0.pdf (accessed July 8, 2025).

[c] COGR. 2025b. Request for Information: "Ensuring Lawful Regulation and Unleashing Innovation to Make America Healthy Again" (Docket No. AHRQ-2025-0001). COGR Response to DHHS Deregulation RFI. https://www.cogr.edu/sites/default/files/Final%20letter%20responding%20to%20HHS%20deregulation%20RFI%20July%202025%20PDF.pdf (accessed July 16, 2025).

[d] HHS. 2024a. *International compilation of human research standards*. https://www.hhs.gov/ohrp/international/compilation-human-research-standards/index.html (accessed July 8, 2025).

[e] *Health Insurance Portability and Accountability Act of 1996*, Public Law No. 104-191, 110 Stat. 1936 (August 21, 1996).

Problem 2: Implementation challenges for the requirement of a single IRB for multisite studies.

When conducting human subjects research, an IRB—an independent committee that reviews research methods and plans to ensure the ethical conduct of human subjects research—must approve the study. In the case of collaborative research conducted across multiple sites, as is common in biomedical research, there could be multiple IRBs duplicating each other's efforts. In 2016, NIH adopted a single IRB policy and in 2018, the Common Rule was also revised to align with the recommendation of the 2016 National Academies report (NASEM, 2016). These changes mandated the use of a single IRB for most federally funded, multisite research (NIH, 2024b).

While the federal single IRB policy for multisite research was intended to eliminate duplicative IRB reviews, it did not fully account for the distinct institutional oversight responsibilities that remain outside the IRB's remit. Institutions retain obligations for compliance and risk management through separate oversight bodies and ancillary committees, such as those overseeing radiation safety, export controls, research security, and scientific review, which IRB review alone cannot replace. As a result, some institutions have their IRBs take on additional compliance responsibilities or conduct separate internal reviews, leading to delays, administrative burden, and variability across institutions. In addition, negotiations to establish a single IRB agreement are complex, and inconsistent implementation across institutions undermines the intended efficiencies of the single IRB policy.

TABLE 2-12 Options to Address Challenges with Implementing a Single IRB (Regulatory Area 6)

Option 6.9: Develop federal guidance clarifying institutional responsibilities under the single Institutional Review Board (IRB) policy

Goal:
Increase clarity on institutional responsibilities under the single IRB policy, reduce duplicative IRB submissions, and streamline multisite research review.

Approach:
Federal agencies, led by the National Institutes of Health (NIH) and U.S. Department of Health and Human Services, could develop and disseminate clear guidance clarifying the distinct roles and responsibilities of single IRBs versus institutional oversight obligations.[a] This guidance could address common issues institutions encounter, such as managing "local context" requirements,[b] reporting mechanisms, and delineating what must remain under institutional purview while minimizing redundant IRB reviews.

Pros:
- Assists institutions in balancing compliance and IRB reliance.
- Reduces unnecessary duplication of reviews while maintaining protections.
- Builds on NIH's existing policy infrastructure and Streamlined, Multisite, Accelerated Resources for Trials (SMART) IRB resources.

Cons:
- Requires coordination across agencies and stakeholder engagement.
- May require adjustments to existing policies to ensure consistency.

Option 6.10: Evaluate and refine the single IRB policy based on implementation data

Goal:
Ensure that the single IRB policy is achieving its intended goals while addressing unintended consequences.

Approach:
Federal agencies could collect and analyze implementation data from institutions and IRBs to evaluate the effectiveness of the single IRB policy, identify barriers, and refine policy requirements where necessary to improve efficiency while maintaining protections. Refinement may include identifying exceptions to requirements as well as considering revisions to submission pathways for exception requests.

Pros:
- Supports evidence-based improvements to the policy.
- Allows for adjustments based on real-world challenges.
- Reinforces trust in federal policy through responsiveness.

Cons:
- Requires systematic data collection and analysis efforts.
- May take time to implement refinements.

continued

TABLE 2-12 Continued

Option 6.11: Encourage adoption of SMART IRB recommendations and local context tools

Goal:
Promote consistent, efficient, and effective implementation of single IRB review across federally funded multisite research while reducing unnecessary duplication and variability.

Approach:
NIH has funded efforts at SMART IRB to develop recommendations and guidelines for harmonizing IRB processes and improving coordination for single IRB review.[c] This option would entail continuing and expanding federal support for the development, refinement, and dissemination of these harmonization guidelines, including tools for managing local context and institutional responsibilities under single IRB frameworks. All federal agencies requiring or supporting single IRB review could actively encourage and support institutions and IRBs in adopting these harmonization guidelines and tools to improve consistency and reduce unnecessary duplication.

Pros:
- Builds on an existing investment in SMART IRB infrastructure.
- Assists institutional IRBs working with a central IRB to improve collaboration.
- Reduces variability and duplication in local context review.
- Encourages consistent interpretation and implementation of single IRB requirements across institutions and agencies.

Cons:
- Additional resources required by federal agencies.

[a] Johnson, A., M. Singleton, J. Ozier, E. Serdoz, J. Beadles, J. Maddox-Regis, S. Mumford, J. Burr, J. Dean, D. Ford, and G. Bernard. 2022. Key lessons and strategies for implementing single IRB review in the Trial Innovation Network. *Journal of Clinical and Translational Science* 6:1–16.

[b] Morain, S. R., J. Bollinger, M. K. Singleton, M. Terkowitz, C. Weston, and J. Sugarman. 2025. Local context review by single institutional review boards: Results from a modified Delphi process. *Journal of Clinical and Translational Science* 9(1):e2.

[c] SMART IRB. n.d. A roadmap to single IRB review. https://smartirb.org/ (accessed July 8, 2025).

Problem 3: Limited flexibility and timeliness within existing regulatory frameworks.

While exempt, expedited, and full board review pathways exist, federal regulations and guidance often lack practical flexibility, clear criteria, or timely processes to adjust oversight proportionally to the level of risk, public health urgency, or societal need. This can lead to unnecessary delays, hindering the rapid launch of lifesaving, rare disease, or other high-priority research when it is most needed.

TABLE 2-13 Options to Address the Limited Flexibility and Timeliness Within Existing Regulatory Frameworks (Regulatory Area 6)

Option 6.12: Establish accelerated and flexible pathways for high-priority and emergency research

Goal:
Enable the rapid and responsible launch of critical high-impact research during emergencies and for high-priority scientific needs.

Approach:
Federal agencies led by the U.S. Department of Health and Human Services (HHS) along with other agencies that conduct or support human subjects research could develop accelerated regulatory pathways and guidance for research addressing life-saving interventions, rare diseases, and urgent public health needs. This could include rolling reviews, time-bound review commitments, prereview consultations, and the ability to activate flexibility through centralized review and streamlined documentation—for example, during public health emergencies or for high-priority research—while ensuring appropriate protections remain in place.

Pros:	Cons:
• Supports timely research that can save lives or address urgent needs.	• Requires clear eligibility criteria to prevent misuse.
• Demonstrates federal commitment to advancing critical science.	• Needs agency coordination and resources for effective implementation.
• Allows for data collection to refine future accelerated pathways.	• Must ensure protections are maintained during expedited processes.

Option 6.13: Establish a federal task force to identify and address bottlenecks in review timelines

Goal:
Reduce systemic delays in the regulatory review processes for all human subjects research.

continued

TABLE 2-13 Continued

Approach:
A dedicated federal task force led by HHS along with other agencies that conduct or support human subjects research could systematically review timelines across agencies to identify inefficiencies, duplicative processes, and other systemic barriers to timely research review. The task force would recommend specific process improvements and, where necessary, regulatory changes to streamline timelines while maintaining participant protections.

Pros:	Cons:
• Provides a structured approach to identifying and addressing inefficiencies. • Encourages collaboration across agencies and with stakeholders. • Can lead to sustainable improvements in review processes across the system.	• Requires interagency leadership and commitment. • Recommendations may require policy or regulatory changes for full implementation.

Problem 4: Inadequate adaptation to evolving research methods and technologies.

Federal regulations have not kept pace with evolving research practices, such as decentralized trials, use of digital health tools, and AI-driven protocols, leading to uncertainty regarding how to ensure compliance while supporting innovation. Though some federal agencies have issued guidance on specific topics, there is no comprehensive cross-agency framework clarifying how human subjects protections apply to emerging research methods and technologies. This creates confusion and inconsistent applications of requirements while potentially slowing the development of innovative participant-centered research approaches.

TABLE 2-14 Option to Address the Inadequate Adaptation to Evolving Research Methods and Technologies (Regulatory Area 6)

Option 6.14: Establish a cross-agency initiative to align and consolidate guidance on emerging research methods

Goal:
Provide clear, consistent, and actionable federal guidance on how existing human subjects protections apply to evolving research methods, reducing confusion and supporting responsible innovation.

TABLE 2-14 Continued

Approach:
The federal government, through the U.S. Department of Health and Human Services and other agencies that conduct or support human subjects research, could establish a cross-agency initiative, led by an oversight body with authority and accountability, to review and consolidate existing agency guidance on emerging research methods and identify areas requiring further clarification. The goal would be to develop a clear, unified federal framework or compendium that aligns interpretations across agencies, fills critical gaps, and removes inconsistencies while preserving participant protections.

Pros:
- Clarifies and aligns existing fragmented guidance across agencies.
- Supports innovation while maintaining participant protections.
- Avoids unnecessary new guidance that adds complexity.
- Provides institutions and Institutional Review Boards with a reliable reference, reducing administrative uncertainty.

Cons:
- Requires sustained interagency coordination.
- May identify areas requiring future regulatory or policy updates.

Problem 5: Lack of federal guidance on integrating nonhuman subjects requirements within human research protection programs.

Federal requirements outside of direct human subjects research oversight, such as data security, research security, export controls, and privacy regulations, often intersect with human subjects research without clear federal guidance on how these requirements should be integrated within a Human Research Protection Program, which includes IRB review processes, investigator responsibilities, and institutional oversight obligations. These intersecting requirements can directly affect protocol design, data collection, data sharing, and international collaboration within human subjects research, yet there is often no clear framework clarifying how institutions should interpret, prioritize, or operationalize these obligations alongside human subjects protections. This lack of integration guidance creates complexity and uncertainty for institutions, leading to duplicative or inconsistent compliance processes, increased administrative burden, and delays in study initiation without necessarily enhancing participant protections.

TABLE 2-15 Options to Address the Lack of Federal Guidance on Integrating Nonhuman Subjects Requirements Within Human Research Protection Programs (Regulatory Area 6)

Option 6.15: Establish a federal integration task force with clear authority and timelines

Goal:
Improve integration of human subjects oversight with nonhuman subjects regulation and increase clarity for institutions.

Approach:
The federal government could establish a cross-agency task force, coordinated by the Office of Science and Technology Policy and Office of Management and Budget, to systematically review the intersection of human subjects research regulations and oversight in data security, research security, export controls, and privacy regulations. The task force could clarify how intersecting nonhuman subjects regulations should integrate with human subjects oversight. It could engage stakeholders throughout the process to ensure practical implementation and to promote alignment with evolving research practices while maintaining participant protections. To ensure effectiveness, the task force could be explicitly charged with recommending and monitoring implementation timelines, with progress reports made public to enhance accountability and transparency.

Pros:
- Promotes consistent, clear, and coordinated federal guidance.
- Increases coordination across multiple areas of regulation and oversight.

Cons:
- Requires sustained interagency leadership and commitment.
- May require regulatory or policy changes for full implementation.
- Coordination across diverse agencies and missions may present challenges.

Option 6.16: Develop a federal interactive compliance integration tool

Goal:
Enable institutions and researchers to determine which intersecting federal requirements apply to their research and how to operationalize them within the Human Research Protection Program (HRPP) processes without unnecessary duplication.

Approach:
Federal agencies led by the U.S. Department of Health and Human Services and other agencies that conduct or support human subjects research could jointly develop an interactive, web-based, or application programming interface-enabled digital tool that helps institutions, Institutional Review Boards, and investigators determine which nonhuman subjects federal requirements regarding research security, export controls, data security, and privacy laws apply to specific human subjects research protocols. This tool would also provide advice on how to integrate them into HRPP workflows efficiently. Ideally, users would enter key project details, such as funding source, study type, data type, and any foreign collaborations, and receive customized output outlining applicable nonhuman subjects research requirements with clear action steps.

TABLE 2-15 Continued

Pros:	Cons:
• Provides clear, accessible, project-specific compliance pathways. • Reduces confusion, delays, and over-implementation of federal requirements. • Increases efficiency while maintaining participant protections and federal compliance. • Scalable to accommodate updates and expansions as federal rules evolve.	• Requires federal investment in development and maintenance. • Agencies must commit to populating and updating guidance consistently. • Complex cases may still require expert interpretation.

REGULATORY AREA 7: RESEARCH USING NONHUMAN ANIMALS

Research using nonhuman animals is governed by an expansive set of requirements and regulations intended to ensure the welfare of vertebrate animals used in research. These protections are important but have unfortunately faced challenges as they have grown in complexity over time in ways that can slow down research without adding significantly to animal welfare. Problems in the current system for regulating animal research include redundancies, contradictions,[26] outdated or overly detailed requirements, and onerous paperwork and reporting that provide little if any benefit for animal welfare.[27]

One of the significant challenges in animal research is the multiagency structure for oversight including USDA Animal and Plant Health Inspection Service, NIH Office of Laboratory Animal Welfare (OLAW), VA, DOD, and others (DOD, 2025; NIH OLAW, 2015, 2024; VA, n.d.). This has created a regulatory framework of conflicting and duplicative require-

---

[26] One such contradiction involves policies regarding cage sizes for lab animals. USDA relies on its own regulations for cage sizes under 9 C.F.R., while PHS relies on minimum cage size recommendations in the *Guide for the Care and Use of Laboratory Animals*, which are not fully aligned with each other.

[27] While the committee was preparing this report, NIH announced on July 18, 2025, that new funding opportunities will prioritize human-based technologies and models and will encourage alternatives to animal models. In addition, FDA has announced plans to reduce or phase out animal testing requirements for certain drugs and biologics.

ments. Minor efforts to harmonize and streamline regulations and policies have occurred. For example, USDA changed its annual review of animal activities by the Institutional Animal Care and Use Committee to align with PHS requirements. However, the changes that have occurred have not eliminated duplicative and at times contradictory oversight of animal research and, unlike human subjects research, animal research has not developed or implemented a Common Rule structure.

Other efforts have been made to address these challenges. In 2019, as required under Title II, Section 2034(d) of the 2016 21st Century Cures Act (NIH et al., 2019), NIH, USDA, and FDA convened a working group to review existing policies and regulations related to the care and use of laboratory animals and make recommendations to reduce administrative burden. The working group concentrated on reducing duplicative regulations and policies and improving coordination across agencies. The report focused mainly on actions to revise the existing structure and "reduce administrative burden on investigators while maintaining the integrity and credibility of research findings and protection of research animals" (FDA, n.d; NIH et al., 2019).

There have been some efforts by outside groups such as the Federal Demonstration Partnership (FDP) to help with streamlining management of regulations for research with nonhuman animals. In 2024, the FDP piloted an online repository called the Compliance Unit Standard Procedures where institutions can share best practices and standard procedures for animal care and welfare. This effort is intended to provide a database repository of consistent and compliant procedures (Bury and Cowell, 2024). The effort is still in early stages and at the time of writing was open only to FDP member institutions.

Inefficient structures and requirements for maintaining Animal Welfare Assurances also increase regulatory workload. Institutions must have a valid Animal Welfare Assurance to conduct PHS-funded research with animals. The current system relies on email correspondence to manage assurances rather than use of a digital platform, as is the case for human subjects research assurances through the HHS Office for Human Research Protections, HHS ORI, and the NIH Office of Science Policy. In addition, OLAW requires a substantially detailed description of institutional animal care and use programs, typically at least 20 pages long but often longer. This can result in processes that take months to review and often involves one or more rounds of revision by the institution.

Finally, challenges arise from unclear and overly strict guidance from OLAW. OLAW, while not a regulatory agency, creates policies, guidance, and recommendations. To this end, OLAW has created more than 150 documents outlining detailed paperwork and administrative requirements as well as guidance for research animal programs that in many instances provide no direct benefit to animal welfare but consume precious research resources. While guidance documents do not carry legal or regulatory power, this fact is often not communicated effectively. As a result, "guidance" documents become requirements imposed on researchers and institutions based on interpretation (COGR, 2017). Therefore, at times, researchers and institutions interpret guidance documents as official regulatory policy and adhere to recommendations that are not required (NASEM, 2016). Instead of a single, clear, and authoritative standard, institutions are left navigating a fragmented and ever-growing body of advisory material, contributing to regulatory creep and administrative burden.

In part, this has occurred because of an overly strict interpretation of the *Guide for the Care and Use of Laboratory Animals* (the *Guide*) (NRC, 2011). The *Guide* serves as a key document for all research with nonhuman animal subjects but was not intended to be interpreted as regulatory requirement. In addition, the *Guide* has not been sufficiently updated since 2011, leading to out-of-date guidance, and no new funding has been allocated for a present-day revision that would capture important advances in technology and improved knowledge of best practices (COGR, 2017).

The committee offers a number of options below to address these challenges, focusing on streamlining, harmonization, and ensuring clear and up-to-date guidance. The committee also provides options to enhance the digital infrastructure of agencies, streamline guidance, and provide regular updates to the *Guide*.

Problem 1: Lack of harmonization in regulation of research using nonhuman animals across federal agencies.

Multiple federal agencies have regulations and requirements governing the use of vertebrate animals in research as noted previously. Each of these agencies have disparate or additive administrative requirements, regulations, and policies that at times conflict with or directly contradict each other, creating challenges for researchers.

TABLE 2-16 Options to Address the Lack of Harmonization in the Regulation of Research Using Nonhuman Animals Across Federal Agencies (Regulatory Area 7)

Option 7.1: Establish a single agency or office to oversee the use of animals in research

Goal:
Create a streamlined set of regulatory standards developed and implemented by a single body, allowing for efficient and expedient compliance with consistency for vertebrate animals used in research regardless of funding source.

Approach:
An act of Congress could create a new government agency or mandate the sole use of one existing agency. This centralized agency would be responsible for developing uniform and consistent regulations, standards, and administrative processes for all U.S. research institutions.

Pros:	Cons:
• Allows the opportunity to review current structures and determine which requirements to keep and develop utilizing scientific evidence. • Coordinates oversight with federal stakeholders more effectively to make quicker, more coherent decisions without needing to reconcile multiple agencies' differing priorities, processes, and paperwork.	• Requires changes to the statutory authority of the U.S. Department of Agriculture (USDA). • Increases burden for institutions and the USDA if statutory authority for the USDA was extended to animal species currently exempted from the Animal Welfare Act, which has not been previously enacted by Congress.

Option 7.2: Eliminate the Animal Care and Use Review Office (ACURO) in the U.S. Department of Defense and establish the USDA as the sole regulatory agency for research of species covered by the Animal Welfare Act and the Office of Laboratory Animal Welfare (OLAW) as the sole oversight body for research funded by the U.S. Department of Health and Human Services involving all other vertebrate (non-USDA covered) species

Goal:
Reduce the number of agencies involved in the regulation of research animals while relying on existing agencies and structures to streamline the process.

TABLE 2-16 Continued

Approach:
ACURO and other similar agency offices could be eliminated as they not only duplicate oversight by the USDA and OLAW but also re-review projects already approved by the federally mandated Institutional Animal Care and Use Committee. With this option, all oversight powers could be contained within USDA and OLAW. The overlapping oversight components for USDA and OLAW could be eliminated, but each group would maintain a different scope of oversight, clearly delineated by which vertebrate species are covered by the Animal Welfare Act and removing duplicative and contradictory compliance requirements for species that are both covered by the USDA and involved in Public Health Service- and U.S. National Science Foundation-funded research.[a]

Pros:	Cons:
• Streamlines current processes and allows for the removal of redundancies and conflicting requirements. • Relies on existing agencies and does not require congressional action.	• Maintains multiple agencies involved in animal research oversight. • Continues the current workload for researchers using multiple species, spanning oversight by two agencies (USDA and the National Institutes of Health's [NIH's] OLAW).

Option 7.3: Increase coordination between agencies to provide for consistency between all federal agencies involved in oversight of animal research

Goal:
Increase interagency coordination and consistency.

Approach:
Improved interagency coordination could be undertaken by establishing a new or empowering an existing National Science and Technology Council subcommittee or working group that includes representatives from USDA, NIH OLAW, ACURO, as well as animal research offices in the U.S. Department of Veterans Affairs, NASA, and National Institute of Standards and Technology.

Pros:	Cons:
• Relies on existing oversight framework. • Lowers costs for agencies and academic institutions with more consistency.	• Maintains multiple agencies involved in regulating the use of animals in research. • Repeats prior efforts under the 21st Century Cures Act that resulted in only a few streamlining actions.

[a] COGR (Council on Governmental Relations). 2025b. Request for Information: "Ensuring Lawful Regulation and Unleashing Innovation to Make America Healthy Again" (Docket No. AHRQ-2025-0001). COGR Response to DHHS Deregulation RFI. https://www.cogr.edu/sites/default/files/Final%20letter%20responding%20to%20HHS%20deregulation%20RFI%20July%202025%20PDF.pdf (accessed July 16, 2025).

Problem 2: Previously identified burdensome and overly detailed NIH OLAW requirements.

OLAW has developed overly detailed administrative and reporting requirements that significantly increase researcher administrative workload through additional paperwork with little direct benefit to animal welfare. PHS Policy allows for Animal Welfare Assurance approval of up to 5 years, yet OLAW requires renewal every 4 years, along with annual reports, documentation of semi-annual reviews, and in instances where a protocol deviation occurs, in-time reports of noncompliance regardless of the degree to which the deviation affects animal welfare (NASEM, 2016). This stems in part from an overly stringent interpretation of the *Guide*, well beyond what was intended by the authors and research community at the time of the latest version's publication (COGR, 2017; NRC, 2011). Further compounding the burden is the over-reliance on email and lack of online submission portals, which are used by other offices within HHS for analogous purposes.

TABLE 2-17 Options to Address Burdensome National Institutes of Health (NIH) Office of Laboratory Animal Welfare (OLAW) Requirements (Regulatory Area 7)

Option 7.4: OLAW can streamline guidance to only the Public Health Service (PHS) Policy and the *Guide for the Care and Use of Laboratory Animals*[a] and interpret the *Guide* as it was intended
Goal: Avoid unnecessary and time-intensive compliance with additional suggestions that are not found in OLAW's foundational documents and do not help researchers or improve the welfare of research animals.
Approach: Use the two required documents, PHS Policy and the *Guide*, as the main guidance from OLAW and eliminate multiple paperwork-based requirements. OLAW could eliminate most, if not all, of its guidance requirements that fall outside the direct scope of the PHS Policy and the *Guide* and, moving forward, could provide institutions the flexibility to interpret and use the *Guide* as it was intended.[b] OLAW could also clarify any guidance that remains does not have legal or regulatory authority.[c]

TABLE 2-17 Continued

Pros:	Cons:
• Increases institutional and researcher flexibility in providing animal care. • Shifts resources from a paperwork focus to an animal-based focus. • Reduces the number of overly detailed OLAW requirements that do not provide direct benefits for research animals.	• If implemented without a process to ensure sufficient and consistent revisions to the *Guide* to keep it up to date, this may introduce challenges for new and emerging practices that are not currently covered under any guidance.

Option 7.5: Update digital infrastructure and Animal Welfare Assurance processes for OLAW in alignment with analogous U.S. Department of Health and Human Services (HHS) oversight offices

Goal:
Use an existing model of assurance and registration approvals to significantly improve the effectiveness and efficiency of the NIH Assurance approval and renewal process.

Approach:
The OLAW Assurance process could be streamlined by aligning it with other HHS office entities that also require assurances or registrations, including HHS Office for Human Research Protections, HHS Office of Research Integrity (ORI), and NIH Office of Science Policy. Not only do the other offices use digital platforms for registration and assurance reviews, they also have streamlined such information to approximately two pages, eliminating the need for dozens of text-heavy pages that must be reviewed and critiqued. OLAW Assurance renewal can take months. In contrast, with a more streamlined process, the renewal process for HHS ORI can be completed within days. Given the successful implementation of such efficient, effective, and time-saving workflows at analogous offices, NIH OLAW can leverage an already existing model of enhanced productivity and consistency.

continued

TABLE 2-17 Continued

Pros:	Cons:
• Creates a user-friendly process leveraging existing HHS models that decreases administrative workload for institutions and NIH OLAW. • Allows NIH OLAW to move to an every-5-year renewal process as provided for in the PHS Policy.	• Costs and coordination efforts across offices would be needed in order to update NIH OLAW's digital infrastructure and to create one submission platform for research compliance registrations and assurances for HHS.

^a NRC (National Research Council). 2011. *Guide for the care and use of laboratory animals: Eighth edition*. Washington, DC: The National Academies Press.

^b COGR (Council on Governmental Relations). 2025b. Request for Information: "Ensuring Lawful Regulation and Unleashing Innovation to Make America Healthy Again" (Docket No. AHRQ-2025-0001). COGR Response to DHHS Deregulation RFI. https://www.cogr.edu/sites/default/files/Final%20letter%20responding%20to%20HHS%20deregulation%20RFI%20July%202025%20PDF.pdf (accessed July 16, 2025).

^c COGR. 2017. *Reforming animal research regulations: Workshop recommendations to reduce regulatory burden*. https://www.cogr.edu/sites/default/files/Animal-Regulatory-Report-October2017.pdf (accessed June 24, 2025).

Problem 3: Lack of a sustainable mechanism for revising the *Guide for the Care and Use of Laboratory Animals*.

The *Guide*, originally published in 1963 and last revised in 2011, is a fundamental document used as guidance throughout the global research community working with laboratory animals. The *Guide* was updated with some regularity following its initial publication in 1963, with NIH issuing revisions in 1965, 1968, 1972, 1978, 1985, 1996, and 2011. Information in the current *Guide* does not reflect the most recent scientific advances, and research practices for many laboratory settings, and does not always offer guidance on nontraditional research applications and animal models. As a result, some requirements in the *Guide* are outdated and may lead to unnecessary efforts, given the availability of new knowledge and best practices. In the absence of updated guidance, NIH OLAW has issued more than 150 commentaries, FAQs, and guidance documents aimed at clarifying expectations. Although these materials are presented as recommendations, unless

TABLE 2-18 Options to Address the Lack of a Sustainable Mechanism for Revising the Guide for the Care and Use of Laboratory Animals (Regulatory Area 7)

Option 7.6: The National Institutes of Health, NASA, U.S. Department of Veterans Affairs, U.S. Department of Defense, and other federal agencies that require extramurally funded institutions to follow the *Guide for the Care and Use of Laboratory Animals* (the *Guide*)[a] would financially sponsor regular revisions

Goal:
Ensure the *Guide* is reflective of current laboratory animal science knowledge and standards.

Approach:
Agencies that require adherence to the *Guide* could financially sponsor revisions to the *Guide* on a predetermined periodic basis.

Pros:	Cons:
• Provides funding for a revision to the *Guide* to ensure guidance is up to date and reflective of current technology and best practices. • Agencies can continue to outsource the task of *Guide* revision.	• Requires cross-agency coordination for revisions to the *Guide* rather than a centralized approach.

Option 7.7: Congress mandates that the *Guide* be updated and revisited periodically

Goal:
Ensure the *Guide* is reflective of current laboratory animal science knowledge and standards.

Approach:
Congress can require that the *Guide* be updated periodically and appropriate funds for the effort.

Pros:	Cons:
• Regular updates to the *Guide* would be required and ensure that the *Guide* is kept up to date and reflective of present needs and situations. • This could come with funds specifically set aside for this effort.	• Greater initial effort required to have this pass Congress.

[a] NRC (National Research Council). 2011. *Guide for the care and use of laboratory animals: Eighth edition.* Washington, DC: The National Academies Press.

specifically mandated, they often function as de facto requirements across academic institutions. NIH OLAW has attempted to provide some flexibility by allowing institutions to use an alternative approach if they satisfy the requirements of the PHS Policy. Revisions of the *Guide* have been supported by funding from NIH and other governmental and nongovernmental entities, but no dedicated funding has been allocated for updates in recent years.

CONCLUDING THOUGHTS

The U.S scientific, engineering, and biomedical enterprise has long been the envy of the world, driving innovation and discovery across sectors that has benefited and shaped society. Ensuring that federally funded science is safe, is conducted with integrity, and protects the interests of the public requires federal oversight and regulation. However, over time, these regulations and requirements have proliferated without sufficient checks, resulting in duplicative agency requirements and disjointed, outdated systems that add heavy workload to implement and hinder innovation. In many cases, a lack of alignment or harmonization, overlapping requirements, and inconsistent implementation have contributed to a fragmented and complex oversight environment, where multiple regulatory demands may apply simultaneously, creating inefficiencies and diverting resources from research itself. A comprehensive reexamination is needed to ensure that regulations and policies continue to support the integrity, security, transparency, and ethical conduct of research while promoting coordination, clarity, and flexibility across the research enterprise.

For years, researchers, groups representing higher education institutions, policymakers, and others have called for reforms to the oversight of federally funded research that provide clear and actionable methods for decreasing burden. However, little progress has been made in implementing these reforms. The U.S. scientific enterprise continues to experience domestic challenges (NASEM, 2023) while competition from other countries is increasing, making it essential that American science advances without unnecessary hindrance. Therefore, it is critical that the federal government take action to revise federal research requirements and oversight processes.

Although the reform options outlined in this report take varying degrees of effort and resources to implement, there is at least one recent example of a potentially promising reform effort that could serve as a model of implementation and demonstrate the value of embracing new approaches. In response to growing concerns about foreign threats to the

OPTIONS TO OPTIMIZE THE RESEARCH ENTERPRISE   101

security and integrity of U.S. research, the CHIPS and Science Act of 2022[28] directed NSF to establish an organization to help researchers meet federal research security requirements. This organization, established in September 2024 and now known as the SECURE Program, including the SECURE Center and SECURE Analytics, serves to connect the research community and collectively design and develop resources and tools to address research security risks and federal agency research security requirements. Although it is relatively new and its evaluation is ongoing, the SECURE Center could be a model for community codesigned resources that facilitate coordinated implementation across institutions. Realization of the SECURE Program stemmed from significant federal-wide interest and concerns from Congress, the White House, national security agencies, and research funding and other agencies and offices about malign foreign influence. Absent this kind of intense focus in other areas, achieving progress and coordination has proved to be incredibly challenging.

Furthermore, leveraging emerging technologies, such as AI, as several options in this chapter propose, may ensure that researchers and regulators both incorporate technology appropriately to increase efficiency. As new technologies such as AI are developed and operationalized, they have the potential to substantially reduce the time, energy, and resources spent across various stages of the research timeline. Special care should be taken to ensure that all institutions have access to these new technologies. Smaller Emerging Research Institutions[29] and traditionally under-resourced institutions (NASFAA, 2021) often struggle with having the necessary resources to utilize technologies such as AI across their institutions. Widening the use of innovations may require targeted funding for infrastructure development to ensure that both small and large institutions can benefit from the potential of burden reduction tools like AI. Integrating novel tools into current and future regulatory activities can significantly reduce the time researchers spend navigating compliance, particularly when coupled with the options for streamlining, harmonization, and reduction detailed previously.

At the same time, as new transformative technologies arise, they will likely require their own regulations and guidance to ensure ethical use in research as well as alignment across agencies. For example, as new technologies such as computational modeling of complex systems, organ-on-a-chip

---

[28] *Creating Helpful Incentives to Produce Semiconductors for America (CHIPS) and Science Act of 2022*, Public Law 117-167 (August 9, 2022).

[29] *Definitions*, 42 U.S.C. § 18901 (August 11, 2025).

systems, and other new approach methodologies are developed, federal agencies are crafting guidance for use of these new technologies and need to coordinate these efforts (FDA, 2025). In addition, developing appropriate AI models requires substantial time and resources. While these technologies have the possibility to produce transformative change, there is still work and cost attached to getting them there.

As various actors work to ensure a more efficient and streamlined regulatory environment, the committee once again encourages policymakers to consider the three principles detailed in this chapter: harmonize research requirements across agencies, take a risk-tiered approach to new requirements, and use technology to simplify requirements and their implementation to the extent possible. Adhering to these principles can ensure an appropriate balance between the oversight needed to ensure federally funded science is safe, ethical, and responsive to the interests of the public,

---

BOX 2-1
Illustrative Case of a Possible Future in the Regulatory Environment

Dr. Linh Tran, an associate professor of robotics at an emerging research university, had always known what she wanted to do. As a child, she watched her sister struggle after losing the use of her arm in an accident. Dr. Tran dreamed of building prosthetics that could restore both motion and dignity—devices as intuitive as a natural limb.

Now, years later, Dr. Tran was on the brink of realizing that vision. Her lab had made major advances in brain-computer interfaces, and her newest prototypes—infused with artificial intelligence—showed the potential to respond to human thought with lifelike precision. With support from the U.S. National Science Foundation (NSF), the U.S. Department of Defense (DOD), and an industry partner, her team was ready to scale. And unlike years past, the process to get there had not been a bureaucratic marathon.

Thanks to recent reforms, NSF and DOD had adopted a shared administrative platform. Dr. Tran submitted her proposal packages to each agency through a unified research portal. To her surprise, both agencies had adopted the same two-step proposal process. Her initial proposals were only five pages long, and upon learning they were to be

and reducing administrative workloads that can liberate and catalyze needed scientific discovery and innovation to advance the well-being, prosperity, and security of the nation.

In Chapter 1, the committee laid out an example of the types of burdens the current research requirements impose on researchers. In Box 2-1, the committee describes the same researcher described in Chapter 1 but in a world where research regulations and requirements have been reformed. This world is a possibility when the principles of harmonization, risk-tiered regulation, and user-friendly technology guide changes to federal research regulation and oversight.

---

funded, she was asked for more detailed documentation. In this process, budget templates, biosketches, and compliance documentation were harmonized, and redundant questions about team composition and institutional approvals were eliminated.

As her projects launched, she completed one annual conflict of interest disclosure as agencies had aligned their definitions of "significant financial interest" and other requirements. At the same time, her postdoc, a brilliant biosystems engineer from South Korea, was cleared to work on the project through a streamlined vetting process for foreign collaborators. Questions about export controls and research security were addressed easily because the federal government had significantly harmonized and clarified its screening criteria.

Dr. Tran's postdoc arrived within months and hit the ground running. DOD no longer required its own Institutional Review Board (IRB) approval, so when the university IRB approved the study, the postdoc was ready to begin recruiting subjects.

By the end of the first year, Dr. Tran's lab had published a major paper, filed a provisional patent, and started to develop plans for a pilot clinical study with volunteers. She still worked long hours, but they were spent mentoring students, analyzing data, and pushing science forward with far fewer hours on administrative requirements. The barriers were lower, the systems were smarter, and Dr. Tran finally felt like the pace of discovery matched the urgency of the need.

REFERENCES

AAMC (Association of American Medical Colleges). 2020. *AAMC Conflicts of Interest Metrics Project - measuring the impact of the public health service regulations on conflicts of interest.* https://www.aamc.org/media/50386/download (accessed July 2, 2025).

ACUS (Administrative Conference of the United States). n.d. *Improving the efficiency of the Paperwork Reduction Act* . https://www.nationalaffairs.com/publications/detail/regulation-beyond-structure-and-process (accessed July 2, 2025).

BIS (Department of Commerce Bureau of Industry and Security). 2011. Deemed exports and fundamental research. https://www.bis.doc.gov/index.php/2011-09-08-19-43-48. (accessed August 12, 2025).

Burr, J. S., A. Johnson, A. Risenmay, S. Bisping, E. S. Serdoz, W. Coleman, et al. 2022. Demonstration project: Transitioning a research network to new single IRB platforms. *Ethics & Human Research*, 44(6): 32–38. https://doi.org/10.1002/eahr.500149.

Bury, S., and A. Cowell. 2024. "Making Compliance Easier: A New Resource for Animal Care & Use Programs." NCURA Magazine, October/November 2024. Washington, DC: Federal Demonstration Partnership. https://thefdp.org/wp-content/uploads/OctNov-2024-NCURA-Mag-CUSP.pdf.

Cathrall, H. 2018. Single IRB review: Does it really decrease administrative burden? *Ampersand*, PRIM&R, December 12, 2018. https://blog.primr.org/single-irb-review-aer18-precon/ (accessed July 2, 2025).

CDC (U.S. Centers for Disease Control and Prevention). 2020. *Biosafety in microbiological and biomedical laboratories 6th edition.* https://www.cdc.gov/labs/pdf/SF__19_308133-A_BMBL6_00-BOOK-WEB-final-3.pdf (accessed July 1, 2025).

COGR (Council on Governmental Relations). 2017. *Reforming animal research regulations: Workshop recommendations to reduce regulatory burden.* https://www.cogr.edu/sites/default/files/Animal-Regulatory-Report-October2017.pdf (accessed June 24, 2025).

COGR. 2019. Science, Security, and Foreign Interference. Presentation at the COGR Science and Security Meeting, June 6. Available at: https://www.cogr.edu/sites/default/files/COGR%206%20June%20FINAL%20-%20%20Read-Only.pdf.

COGR. 2021. *Principles for evaluating conflict of commitment concerns in academic research.* https://www.cogr.edu/sites/default/files/Final%20for%20publication%20COC%20Principles%20Document%20V%202%20Sept%202021%202021.pdf (accessed July 15, 2025).

COGR. 2022. *Research security and the cost of compliance phase / report.* https://www.cogr.edu/sites/default/files/Version%20Dec%205%202022%20research%20security%20costs%20survey%20FINAL.pdf (accessed July 15, 2025).

COGR. 2025a. *Actionable ideas to improve government efficiency affecting the performance of research.* https://www.cogr.edu/sites/default/files/Actionable Ideas to Improve Gov Efficiency COGR_0.pdf (accessed July 8, 2025).

COGR. 2025b. Request for Information: "Ensuring Lawful Regulation and Unleashing Innovation to Make American Healthy Again" (Docket No. AHRQ-2025-0001). COGR Response to DHHS Deregulation RFI. https://www.cogr.edu/sites/default/files/Final%20letter%20responding%20to%20HHS%20deregulation%20RFI%20July%202025%20PDF.pdf (accessed July 16, 2025).

Crowell & Moring LLP. 2023. "New U.S. Department of Defense Policy Imposes Security Reviews for Universities and Labs Engaging in Fundamental Research." September 14. Available at: www.crowell.com/en/insights/client-alerts/new-us-department-of-defense-policy-imposes-security-reviews-for-universities-and-labs-engaging-in-fundamental-research.

CRS (Congressional Research Service). 2019. *The U.S. export control system and the Export Control Reform Initiative.* https://www.congress.gov/crs_external_products/R/PDF/R41916/R41916.46.pdf (accessed July 1, 2025).

CRS. 2020. *The U.S. export control system and the Export Control Reform Initiative.* https://sgp.fas.org/crs/natsec/R41916.pdf (accessed August 5, 2025).

CRS. 2025. *Oversight of gain-of-function research with pathogens: Issues for Congress.* https://www.congress.gov/crs-product/R47114 (accessed July 11, 2025).

Decrappeo, A., D. Kennedy, J. Trapani, and T. Smith. 2011. Reforming regulation of research universities. *Issues in Science and Technology* 27(4). https://issues.org/smith-5/ (accessed August 7, 2025).

Defino, T. 2025. ORI makes first misconduct finding since October; concerns agency is "slowing down." *Healthcare Compliance Association* 12(4). https://compliancecosmos.org/ori-makes-first-misconduct-finding-october-concerns-agency-slowing-down.

DOD (U.S. Department of Defense). 2025. *Regulations, standards, and requirements.* https://mrdc.health.mil/index.cfm/collaborate/research_protections/acuro/regulations (accessed June 24, 2025).

DOE (Department of Energy). 2021. Financial Assistance Letter No. FAL 2022-02: Interim Conflict of Interest Policy for Financial Assistance. December 20. Available at: https://www.energy.gov/sites/default/files/2021-12/Interim%20COI%20Policy%20FAL2022-02%20to%20SPEs.pdf (accessed September 12, 2025).

DOS (U.S. Department of State). 2013. *The president's Export Control Reform Initiative: Reinventing the system and promoting national security.* https://2009-2017.state.gov/r/pa/pl/2013/209319.htm (accessed August 5, 2025).

Ehidiamen, A. J., and O. O. Oladapo. 2024. Enhancing ethical standards in clinical trials: A deep dive into regulatory compliance, informed consent, and participant rights protection frameworks. *World Journal of Biology Pharmacy and Health Sciences* 20(01): 309–320.

EPA (Environmental Protection Agency). 2025. About risk assessment. https://www.epa.gov/risk/about-risk-assessment#whatisrisk (accessed August 12, 2025).

eRA (Electronic Research Administration). 2025. *eRA Commons user guide.* https://www.era.nih.gov/docs/Commons_UserGuide.pdf (accessed August 13, 2025).

FDA (U.S. Food and Drug Administration). n.d. 21st century cures act. https://www.fda.gov/regulatory-information/selected-amendments-fdc-act/21st-century-cures-act (accessed July 15, 2025).

FDA. 2013. *Guidance for clinical investigators, industry, and FDA staff: Financial disclosure by clinical investigators.* https://www.fda.gov/media/85293/download (accessed July 14, 2025).

FDA. 2025. *FDA announces plan to phase out animal testing requirement for monoclonal antibodies and other drugs.* https://www.fda.gov/news-events/press-announcements/fda-announces-plan-phase-out-animal-testing-requirement-monoclonal-antibodies-and-other-drugs (accessed July 22, 2025).

FDP (Federal Demonstration Partnership). n.d. *Who we are.* https://thefdp.org/ (accessed July 11, 2025).

FDP. 2025. *Organization history.* https://thefdp.org/organization/history/#tab-id-2 (accessed July 11, 2025).

FEMA (Federal Emergency Management Agency). 2021. *Information update about the export allocation rule on medical supplies and equipment for COVID-19.* https://www.fema.gov/fact-sheet/allocation-rule-personal-protective-equipment-exports (accessed July 1, 2025).

Fleischman, A. 2005. Regulating research with human subjects—is the system broken? *Transactions of the American Clinical and Climatological Association* 116: 91–102.

GAO (U.S. Government Accountability Office). 2023. *Institutional review boards: Actions needed to improve federal oversight and examine effectiveness.* https://www.gao.gov/assets/gao-23-104721.pdf (accessed July 15, 2025).

GAO. 2024. *Grant management: Action needed to ensure consistency and usefulness of new data standards.* https://www.gao.gov/products/gao-24-106164 (accessed July 11, 2025).

Green, J., P. Goodman, A. Kirby, N. Cobb, and B. E. Bierer. 2023. Implementation of single IRB review for multisite human subjects research: Persistent challenges and possible solutions. *Journal of Clinical and Translational Science* 7(1): e99. https://doi.org/10.1017/cts.2023.517.

HHS (U.S Department of Health and Human Services). 2017. *Revised common rule.* https://www.hhs.gov/ohrp/regulations-and-policy/regulations/finalized-revisions-common-rule/index.html (accessed July 8, 2025).

HHS. 2018. *The Belmont Report.* https://www.hhs.gov/ohrp/regulations-and-policy/belmont-report/read-the-belmont-report/index.html (accessed July 8, 2025).

HHS. 2024a. *International compilation of human research standards.* https://www.hhs.gov/ohrp/international/compilation-human-research-standards/index.html (accessed July 8, 2025).

HHS. 2024b. *The Belmont Report.* https://www.hhs.gov/ohrp/regulations-and-policy/belmont-report/index.html (accessed July 8, 2025).

HHS. 2025. *Federal policy for the protection of human subjects ("Common Rule").* https://www.hhs.gov/ohrp/regulations-and-policy/regulations/common-rule/index.html (accessed July 8, 2025).

HHS, CDC (U.S. Centers for Disease Control and Prevention), and NIH (National Institutes of Health). 2020. Biosafety in microbiological and biomedical laboratories (BMBL), 6th Edition. https://www.cdc.gov/labs/pdf/SF__19_308133-A_BMBL6_00-BOOK-WEB-final-3.pdf. (accessed August 12, 2025).

HHS and USDA (U.S. Department of Agriculture). n.d. Federal select agent program. https://www.selectagents.gov/ (accessed August 8, 2025).

Insinna, V. 2017. Defense industry hopeful Trump will pick up Obama's legacy of export control reform. *Defense News*, January 20, 2017. https://www.defensenews.com/air/2017/01/20/defense-industry-hopeful-trump-will-pick-up-obama-s-legacy-of-export-control-reform (accessed August 5, 2025).

Johnson, A., M. Singleton, J. Ozier, E. Serdoz, J. Beadles, J. Maddox-Regis, S. Mumford, J. Burr, J. Dean, D. Ford, and G. Bernard. 2022. Key lessons and strategies for implementing single IRB review in the Trial Innovation Network. *Journal of Clinical and Translational Science* 6:1–16.

Johnson, C. M., and K. M. Dobos. 2019. The evolving landscape of institutional biosafety committees and biosafety programs: Results from a national survey on organizational structure, resources, and practices. *Applied Biosafety: Journal of the American Biological Safety Association* 24(4): 213–219.

Kiritz, J., et al. 2019. Model Risk Tiering: An Exploration of Industry Practices and Principles. *Journal of Risk Management in Financial Institutions* 12(4): 388–399.

Morain, S. R., J. Bollinger, M. K. Singleton, M. Terkowitz, C. Weston, and J. Sugarman. 2025. Local context review by single institutional review boards: Results from a modified Delphi process. *Journal of Clinical and Translational Science* 9(1):e2.

NASA (National Aeronautics and Space Administration). 2023. *Grant Information Circular (GIC) 23-07: Conflict of Interest Policy*. 21 Aug. 2023, effective 1 Dec. 2023. https://www.nasa.gov/wp-content/uploads/2023/09/gic-23-07-conflict-of-interest-policy-0.pdf (accessed September 12, 2025).

NASEM (National Academies of Sciences, Engineering, and Medicine). 2009. *Beyond 'Fortress America': National security controls on science and technology in a globalized world*. Washington, DC: The National Academies Press.

NASEM. 2016. *Optimizing the nation's investment in academic research: A new regulatory framework for the 21st century*. Washington, DC: The National Academies Press.

NASEM. 2022. *Protecting U.S. technological advantage*. Washington, DC: The National Academies Press.

NASEM. 2023. *Transforming research and higher education institutions in the next 75 years: Proceedings of the 2022 Endless Frontier Symposium*. Washington, DC: The National Academies Press.

NASFAA (National Association of Student Financial Aid Administrators). 2021. *Under-resourced schools*. Thought Force Report. https://www.nasfaa.org/uploads/documents/Under-Resourced_Schools_Thought_Force_Report.pdf. (accessed August 12, 2025).

NIH (National Institutes of Health). 2024a. *Financial conflict of interest*. https://grants.nih.gov/policy-and-compliance/policy-topics/fcoi (accessed July 14, 2025).

NIH. 2024b. *Single IRB for multi-site or cooperative research*. https://grants.nih.gov/policy-and-compliance/policy-topics/human-subjects/single-irb-policy-multi-site-research (accessed July 8, 2025).

NIH. 2024c. NIH guidelines for research involving recombinant or synthetic nucleic acid molecules. Office of Science Policy, U.S. Department of Health and Human Services. (Amendments effective September 30, 2024).

NIH. 2025. SciENcv background. https://www.ncbi.nlm.nih.gov/sciencv/background/?hss_channel=lcp-9398777 (accessed August 11, 2025).

NIH OER (National Institutes of Health, Office of Extramural Research). 2024. *Research Project Grants and Other Mechanisms: Competing Applications, Awards, Success Rates, and Funding, by Institute/Center, Mechanism/Funding Source, and Activity Code*. Research Portfolio Online Reporting Tools (RePORT). https://report.nih.gov/funding/nih-budget-and-spending-data-past-fiscal-years/success-rates (accessed August 11, 2025).

NIH OLAW (National Institutes of Health Office of Laboratory Animal Welfare). 2015. *Public Health Service policy on humane care and use of laboratory animals*. https://olaw.nih.gov/policies-laws/phs-policy.htm (accessed June 24, 2025).

NIH OLAW. 2024. *Office of laboratory animal welfare*. https://olaw.nih.gov/home.htm (accessed June 24, 2025).

NIH, USDA (United States Department of Agriculture), and FDA. 2019. *Reducing administrative burden for researchers: Animal care and use in research*. https://olaw.nih.gov/sites/default/files/21CCA_final_report.pdf (accessed June 19, 2025).

NRC (National Research Council). 2011. *Guide for the care and use of laboratory animals: Eighth edition*. Washington, DC: The National Academies Press.

NSF (U.S. National Science Foundation). n.d.-a. *Conflicts of interest*. https://www.nsf.gov/policies/conflict-of-interest (accessed July 14, 2025).

NSF. n.d.-b. *NSPM-33 implementation guidance*. https://www.nsf.gov/bfa/dias/policy/nspm-33-implementation-guidance (accessed July 15, 2025).

NSF. 2023. *NSF 23-613: Research security and integrity information sharing analysis organization (RSI-ISAO)*. https://www.nsf.gov/funding/opportunities/rsi-isao-research-security-integrity-information-sharing-analysis/506033/nsf23-613/solicitation (accessed July 23, 2025).

NSF. 2024. *NSF-backed secure center will support research security, international collaboration*. https://www.nsf.gov/news/nsf-backed-secure-center-will-support-research (accessed July 16, 2025).

NSF. 2025. *FY 2025 Budget Request to Congress: NSF Funding Profile*. Available at: https://nsf-gov-resources.nsf.gov/files/04_fy2025.pdf.

NSTC (National Science and Technology Council). 2022. *Guidance for implementing National Security Presidential Memorandum-33 (NSPM-33) on national security strategy for United States government supported research and development*. https://bidenwhitehouse.archives.gov/wp-content/uploads/2022/01/010422-NSPM-33-Implementation-Guidance.pdf (accessed August 5, 2025).

ORI (The Office of Research Integrity). n.d. *Conflicts of commitment*. https://ori.hhs.gov/education/products/rcradmin/topics/coi/tutorial_4.shtml (accessed July 16, 2025).

ORI. 2000. *Federal research misconduct policy*. https://ori.hhs.gov/federal-research-misconduct-policy (accessed July 11, 2025).

ORI. 2024. *ORI final rule*. https://ori.hhs.gov/blog/ori-final-rule (accessed June 12, 2025).

ORI. 2025. *ORI's first phase of final rule guidance documents released*. https://ori.hhs.gov/blog/oris-first-phase-final-rule-guidance-documents-released (accessed July 11, 2025).

Phillips, T., and J. Earl. 2025. Reflections on the 2024 final rule on Public Health Service policies on research misconduct. *Accountability in Research* 32(5):675–692.

Rudalevige, A. 2018. Regulation beyond structure and process. *National Affairs*. https://www.nationalaffairs.com/publications/detail/regulation-beyond-structure-and-process.

SMART IRB. n.d. *A roadmap to single IRB review*. https://smartirb.org/ (accessed July 8, 2025).

Stricker, A., and D. Albright. 2017. *U.S. export control reform: Impacts and implications*. Washington, DC: Institute for Science and International Security. https://isis-online.org/isis-reports/u.s.-export-control-reform-impacts-and-implications (accessed August 5, 2025).

The White House. 1985. *National policy on the transfer of scientific, technical, and engineering information*. https://www.acq.osd.mil/dpap/dars/pgi/docs/National_Security_Decision_Directive_189.pdf (accessed July 11, 2025).

The White House. 2013. *Fact sheet: implementation of export control reform*. https://obamawhitehouse.archives.gov/the-press-office/2013/03/08/fact-sheet-implementation-export-control-reform. (accessed August 12, 2025).

The White House. 2025. *Improving the safety and security of biological research*. https://www.whitehouse.gov/presidential-actions/2025/05/improving-the-safety-and-security-of-biological-research/ (accessed July 11, 2025).

VA (U.S. Department of Veterans Affairs). n.d. *Oversight and guidelines for animal research in VA*. https://www.research.va.gov/programs/animal_research/overview.cfm (accessed June 24, 2025).

VA and VHA (Veterans Health Administration). 2017. *The office of research oversight (VHA directive 1058)*. https://navao.org/wp-content/uploads/2017/04/VHA-Directive-1058-Office-of-Research-Oversight-3-28-17.pdf (accessed July 11, 2025).

Walch-Patterson, A. 2020. Exemptions and limited institutional review board review: A practical look at the 2018 common rule requirements for exempt research. *Ochsner Journal* 20: 87–94.

Appendix A

Public Meeting Agendas

COMMITTEE MEETING 2
Wednesday, May 21, 2025

OPEN SESSION

10:00–11:00 **Discussion with Sponsors**
Marcia McNutt, President, NAS
David Spergel, President, Simons Foundation

11:00–11:15 **Break**

11:15–11:45 **OSTP Perspectives**
Lynne Parker, Principal Deputy Director, White House Office of Science and Technology Policy

11:45–12:45 **Lunch**

12:45–1:45 **Professional Societies/Advocacy Groups**
Matt Owens, President, COGR
Michele Masucci, Vice Chancellor for Research and Economic Development, University of Maryland System

Alexandra Albinak, Associate Vice Provost for Research Administration, Johns Hopkins University; Co-Chair, Executive Committee, Federal Demonstration Partnership (FDP)

1:45–3:00 **Barriers and Challenges in Research Regulations**
Moderator: Stuart Shapiro, Dean and Professor, Bloustein School of Planning and Public Policy, Rutgers University
Susan Garfinkel, Consultant Owner, Research Integrity Partners; Former Senior Advisor to the Director, Office of Research Integrity
Roger Nober, Director, Regulatory Studies Center, George Washington University
Susan Sedwick, Senior Consulting Specialist, Attain Partners

3:00–3:30 **Break**

Friday, June 6, 2025 (Eastern Standard Time)

OPEN SESSION

3:00–4:00 **Conversation with Federal Research Compliance Officers (Members of the Interagency Working Group for the Common Rule)**
Anne Andrews, Director, Research Protections Office, NIST Scientific Integrity Officer, National Institute of Standards and Technology (NIST)
Stephanie Bruce, Director, Office for Human Research Protections and Bioethics, Office of the Under Secretary for Research and Engineering, Department of Defense (DOD)
Natalie Klein, Acting Director, Department of Health and Human Services Office for Human Research Protections (OHRP)

4:00–5:00 **Operation Warp Speed: Lessons Learned**
Kevin Bugin, Head of Global Regulatory Policy and Intelligence, Amgen; Former Deputy Director of Operations, Office of New Drugs, FDA; Former Chief of Staff, "Operation Warp Speed"

Appendix B

Committee Biographical Sketches

Alan Leshner (NAM) (*Chair*) is CEO, emeritus, of the American Association for the Advancement of Science (AAAS) and former executive publisher of the journal *Science* and the Science family of journals. He served as permanent CEO from December 2001 through February 2015, and then as interim CEO from July to December 2019. Before joining AAAS, Dr. Leshner was director of the National Institute on Drug Abuse at the National Institutes of Health (NIH). He also served as deputy director and acting director of the National Institute of Mental Health and in several roles at the U.S. National Science Foundation (NSF). Before joining the government, Dr. Leshner was professor of psychology at Bucknell University, where he taught and conducted research on the relationship between hormones and behavior. Dr. Leshner is an elected fellow of AAAS, the American Academy of Arts and Sciences, the National Academy of Public Administration, and many others. He is a member and served as vice chair of the Governing Council of the National Academy of Medicine (formerly the Institute of Medicine) of the National Academies of Sciences, Engineering, and Medicine. He served two terms on the National Science Board, appointed first by President Bush and then reappointed by President Obama. Dr. Leshner received PhD and MS degrees in physiological psychology from Rutgers University and an AB in psychology from Franklin and Marshall College. He has received many honors and awards, including the Walsh McDermott Medal from the National Academy of Medicine and seven honorary doctor of science degrees.

David Apatoff is a lawyer specializing in research and development projects, grants, and contracts for research universities and nonprofits. He was a senior partner at the law firm of Arnold & Porter, where he served on the Policy Committee managing the firm's global operations and headed the firm's intellectual property practice. His projects have ranged from biotechnology patent arbitrations and telecommunications infrastructure transactions to information technology licensing agreements and environmental disputes. His primary legal expertise is in the area of government contracts and grants, where he has handled a wide variety of issues including audits, investigations (both civil and criminal), congressional inquiries, bid protests, change orders, performance disputes, cost accounting matters, and claims. He has litigated government contract cases in federal district courts and in the Court of Federal Claims. He has represented several Fortune 100 companies but developed a subspecialty representing nonprofits and research universities. He is a graduate of the University of Chicago Law School.

Linda Coleman is the associate vice provost for research policy and integrity at Stanford University, where she develops and implements strategies to manage compliance risks, strengthen research programs, and ensure adherence to evolving regulatory requirements. She oversees key areas including research security, export controls and global engagement review, conflict of interest and commitment, data governance and privacy, and responsible and ethical conduct of research. Before joining Stanford, Ms. Coleman was director of the Human Research Protection Program at Yale University. In this capacity, she oversaw the Institutional Review Board (IRB) and several non-IRB committees, such as the Radiation Drug Research Committee, Radiation Drug Investigation Committee, Institutional Conflict of Interest Committee (in collaboration with the Conflict of Interest Office), and the Embryonic Stem Cell Research Oversight Committee. Prior to Yale, she held progressive leadership roles, including vice president of legal and regulatory affairs and director of regulatory affairs and general counsel at Quorum Review/Kinetiq (now part of Advarra), an independent IRB and consulting firm serving institutional, independent, and international research sites. Earlier in her career, she was an attorney at Bennett, Bigelow & Leedom, specializing in general health law matters and employment law. Ms. Coleman has contributed to a number of national and international initiatives and served on a range of expert committees. Her work in these areas has included advising on policy development, shaping guidance, and offering strategic input on scalable approaches that support both regulatory compliance and research excellence.

Kelvin Droegemeier is professor of climate, meteorology, and atmospheric sciences and special advisor to the chancellor for science and policy at the University of Illinois Urbana-Champaign. He previously spent 38 years on the faculty at the University of Oklahoma, where he served for nearly a decade as vice president for research. He co-founded and directed one of the first 11 NSF Science and Technology Centers and co-founded an NSF Engineering Research Center. Dr. Droegemeier served as Oklahoma Cabinet secretary for science and technology as well as two terms on the National Science Board, the last 4 years as vice chairman. Most recently, he served as director of the White House Office of Science and Technology Policy (OSTP), science advisor to the president, and acting director of NSF. He is a fellow of the American Meteorological Society and AAAS and has served on and chaired numerous boards, including as chair of the University Corporation for Atmospheric Research and Southeastern Universities Research Association boards of trustees, and presently serves on the Board on Research Data and Information and the Committee on Science, Engineering, Medicine, and Public Policy of the National Academies. In 2023, Dr. Droegemeier authored a book titled *Demystifying the Academic Research Enterprise*.

Melanie Graham is a professor in the Department of Surgery at the University of Minnesota, where she serves as executive vice chair for research strategy and operations and vice chair for faculty development. She also directs the Preclinical Research Center and holds an adjunct appointment in veterinary population medicine. Dr. Graham is also faculty in the Institute for Diabetes, Obesity, and Metabolism and the Stem Cell Institute, and serves as graduate faculty across multiple programs. She trained in experimental surgery and epidemiology at the University of Minnesota and earned her PhD in animal modeling and welfare from the University of Utrecht. A primatologist by training, Dr. Graham's research focuses on translational models for diabetes, immunotherapy, and cell- and gene-based therapies. Her program is supported by funding from federal agencies, foundations, state initiatives, and industry collaborations. Her work advances both scientific innovation and animal welfare, with the goal of delivering transformative therapies to patients in need. She holds the prestigious Goodale Chair in Minimally Invasive Surgery, a recognition of her contributions to both surgical innovation and the field. Dr. Graham is nationally and internationally recognized for her leadership in research practice and policy. She has served on the National Academies' Committee on Nonhuman Primate Model Systems and the UK Medical Research Council's Scientific Landscape Review advisory group. She has held leadership roles on public and private Institutional Animal Care and Use Committees (IACUCs), the Academy of Surgical Research, and the 3Rs Collaborative. She has also contributed to several NIH advisory committees, including those for Fostering Rigorous Research: Lessons Learned from Nonhuman Primate Models, the Immunobiology of Xenotransplantation Program, and the National Institute of Allergy and Infectious Diseases Nonhuman Primate Transplantation Tolerance Cooperative Study Group. In addition, she regularly serves on NIH and other peer review panels and participated in U.S. Food and Drug Administration-sponsored scientific expert workshops.

Lisa Nichols is executive director of research security at the University of Notre Dame where she has oversight for research security, export controls, regulated data/controlled unclassified information, and facility security/classified research. She previously held roles at the NIH, NSF, OSTP, and Council on Governmental Relations (COGR) as well as the University of Pennsylvania and University of Michigan. At NSF, Dr. Nichols led the development of the report *Reducing Investigators' Administrative Workload for Federally Funded Research* under the direction of the National Science Board's Task Force on Administrative Burdens. At COGR she engaged with federal agencies across all compliance areas on behalf of institutions of higher education with a focus on regulatory reform as the director, Research and Regulatory Reform, and at OSTP as assistant director for Academic Engagement where she focused on research regulatory reform and agency coordination of emerging policy, including the development of National Security Presidential Memorandum-33. She served as principal investigator (PI) on an NSF cooperative agreement to develop research security training for the U.S. research community and currently serves as PI on a Safeguarding the Entire Community in the U.S. Research Ecosystem (SECURE) Center subaward as part of its national team to develop research security resources and tools using a community-centered approach. She is co-chair of the Federal Demonstration Partnership's Research Security Subcommittee, consisting of both federal and institutional members, and a member of a higher education association science and security working group and the National Academies' Assessing Research Security Efforts in Higher

Education Workshop Planning Committee. Dr. Nichols holds a PhD in neuroscience from Purdue University and is a former AAAS Science and Technology Policy fellow.

Julia Phillips retired from Sandia National Laboratories in 2015. She culminated her Sandia career by serving as vice president and chief technology officer. Earlier, she spent 14 years at AT&T Bell Laboratories where she performed leading-edge research in thin film epitaxial electronic materials and complex oxides. Dr. Phillips is a member and past home secretary of the National Academy of Engineering and fellow of the American Academy of Arts and Sciences, Materials Research Society (past president), AAAS, and the American Physical Society (APS). She received the George E. Pake Prize from APS "for her leadership and pioneering research in materials physics for industrial and national security applications." She received a BS degree in physics from William and Mary and a PhD in applied physics from Yale University. She currently advises the federal government and research organizations, communicates widely about science and engineering and its national importance, and mentors individuals and groups at diverse career and life stages. Dr. Phillips chairs the Division on Engineering and Physical Sciences Committee of the National Academies (2020–present) and is a member of the National Science Board (2016–2028). While on the National Science Board she has chaired the Committee on Science and Engineering Policy for 6 years and served as the vice chair of the Commission on Merit Review (2023–2025) and as vice chair of the Committee on Awards and Facilities.

Stacy Pritt serves as the associate vice chancellor and chief research compliance officer at the Texas A&M University System. In this position, she establishes research compliance policies and initiatives for research with animals, human participants, and biohazards, along with financial conflict of interest in research and research misconduct for 11 academic institutions and 8 state agencies. Previously, she worked in research compliance in industry and academia including the University of Texas Southwestern Medical Center and Harvard Medical School. During her career, Dr. Pritt has given more than 100 professional lectures and authored or co-authored dozens of publications on the topics of research compliance, security, and administration as well as process improvement, research reproducibility, training, and management. She is the recipient of numerous awards for her training, speaking, and management activities including the American Association for Laboratory Animal Science George R. Collins Award, TurnKey Leader of the Year Award, and Washington State University's College of Veterinary Medicine Outstanding Service Alumni Award. In 2021, she was named as a Distinguished Faculty member for the Society of Research Administrators International. Dr. Pritt earned her BS degree in biology from the California State Polytechnic University at Pomona and her Doctor of Veterinary Medicine degree from Washington State University. She has also earned additional degrees in business and management. She is board certified by the American College of Animal Welfare and is a Certified Professional in IACUC Administration. She has served in numerous leadership positions including vice president of the American Veterinary Medical Association, president of the American College of Animal Welfare, and president of the Laboratory Animal Welfare Training Exchange.

Stuart Shapiro is the dean of the Bloustein School of Planning and Public Policy at Rutgers University. He has been a professor at the Bloustein School since 2003. Prior to that he worked at the federal Office of Information and Regulatory Affairs in the Office of Management and Budget from 1998 to 2003. Dean Shapiro is a nationally recognized expert on the federal regulatory process, the use of cost-benefit analysis in regulatory decision-making, and the Paperwork Reduction Act. He has written 4 books and more than 40 articles on these subjects. He received his PhD in public policy. In 2016, Dean Shapiro served on the National Academies committee that produced the report *Optimizing the Nation's Investment in Academic Research: A New Regulatory Framework for the 21st Century*.

Christopher Viggiani is the associate vice president for research integrity at Oregon State University (OSU). Dr. Viggiani leads the Office of Research Integrity, integrating high ethical and professional standards into OSU's research and innovation enterprise. He oversees programs on human research protections, animal welfare, export controls and international compliance, responsible research practices and misconduct, research security, conflicts of interest, biosafety, and more. Previously, he oversaw university policies and standards at OSU, managing university-wide policy development. Prior to joining OSU, Dr. Viggiani ran the biosafety and biosecurity policy program at NIH. Within the NIH Director's Office of Science Policy, he worked with federal partners across the U.S. government to develop and implement federal policies that promote global health security and advance emerging biotechnologies. He served as executive director of the National Science Advisory Board for Biosecurity and helped lead the U.S. government's deliberative process on dual-use research and gain-of-function studies involving pandemic pathogens. Dr. Viggiani earned a BS from Virginia Tech and a PhD in molecular biology from the University of Southern California and was a postdoctoral fellow at the Johns Hopkins University School of Medicine. His research focused on DNA replication, chromosome stability, and telomeres.

Emanuel Waddell currently serves as professor and chair of the Department of Nanoengineering at North Carolina Agricultural and Technical State University. Previously, he was a program officer at NSF (2019–2022) and associate dean of the College of Science at the University of Alabama in Huntsville (2015–2019). Dr. Waddell's research focuses on analytical chemistry and materials science, with expertise in surface modification techniques for microfluidic systems and polymer substrates. His work has resulted in multiple patents in chemical modification of substrates and surface charge modification within polymer microchannels. He has secured over $1.5 million in competitive grants from agencies including NSF, NASA, and the Camille and Henry Dreyfus Foundation. He received the NSF Director's Award for Achievement in Equal Opportunity/Diversity and Inclusion (2021) and served as president of the National Organization for the Professional Advancement of Black Chemists and Chemical Engineers (2017–2019). He is also a life member of the Alabama Academy of Science. Dr. Waddell earned his PhD in analytical chemistry from Louisiana State University, MS in physical chemistry from the University of Rochester, and BS in chemistry and physics from Morehouse College, with postdoctoral training at the National Institute of Standards and Technology.

Stephen Willard is currently the CEO of ICaPath, Inc., a biotechnology company focused on curing cancer through immunostimulation. He has spent more than 30 years leading biotechnology and pharmaceutical companies in both the United States and France. His experience includes government grant funding, successful IPOs (initial public offerings) and financial market transactions in the hundreds of millions of dollars. He has been the CEO of both private and public companies in the biopharmaceutical sector. He has practiced law in New York, London, and Washington, DC, and was an investment banker for a time. He has developed, executed, and managed multinational partnerships and corporate transactions worldwide. Mr. Willard was a member of the board of directors of E*TRADE Financial and/or its bank from 2000 to 2014, where he had stints as chairman of the Bank Audit Committee, head of the Risk Oversight Committee, and vice chairman. He is highly experienced with boards of directors on both the management and outside board member roles. He has a background in financial institutions, having served as associate director of resolutions for the Federal Deposit Insurance Corporation, a Senior Executive Service-2-level position, from 1991 to 1994. He managed the resolution of the nation's largest troubled banks during this period. His decisions were favorably reviewed in multiple U.S. Government Accountability Office audits. Mr. Willard is a member of the National Science Board's class of 2018–2024.